ZERO TO ONE

Making Our Way

Toward a Conscious Civilization

Bruce 'Zen' Benefiel, DD, MA, MBA

You might think this is going to be about computer code, the binary system of ones and zeros, and how technology is going to help manage our planetary civilization. In a sense you'd be right, but not in the way you might think. Choice is also binary.

In another sense, and perhaps quite literally, the internet is an interface into such a system, driven by both emotion and logic, with the capability to engage people, places and things for various purposes. I perceive there are even greater things we can do toward peace with the technology. We just haven't made the choice to do so, yet.

There is another binary connection that, through the choice of 'yes' or 'no,' molds our reality on a moment-to-moment basis. It, too, is an emotion and logic driven system. The two are nearly inseparable and often in conflict. We all experience that to a greater or lesser degree. It's called conscience.

To the latter, a little background might help you understand the mind behind the meanderings. I was orphaned at birth, told of my adoption at four and a half when my adopted sister arrived. It sent me inward with questions, deep ones for such a young boy especially. Those questions were about family, both on Earth and in Heaven. An inner connection with a 'voice' began around my fifth birthday.

My exploration of self-discovery continued; introduced to out-of-body experiences (OBE) at 7, trips to an orange cigar-shaped cloud from 8-10 shifted to more physical world activity as I entered DeMolay at 13 and achieved Master Councilor at 15. My adopted father was a 32nd Degree Mason, so I grew up with very strong moral and ethical codes. I hold them still today, maybe even more so as the practice of them has imbued my life with amazing and profound experiences.

We ask questions and ponder paths, imagining possibilities for creating a sustainable world and perhaps even our part.

Current quantum theory indicates we have more capacity for becoming self-aware and managing our reality better; individually influencing a collective consciousness.

On the theme of destiny or fate, or even just making better choices, does synchronicity have anything to do with it? It's been said we live half inside and half outside. How do we access a clearer path toward self-actualization and self-realization; a bridge of inner and outer that benefits us all?

If we consider the notion of creating our reality, does that eliminate determinism or is it complementary? How specific is the perfected form, fit and function as a prudent path in the world and are we capable of discovery?

Does this commitment take free will away or refine it with discernment and wisdom? Will knowing more have an effect on our activity in the future? Is your shift in awareness going to impact the world around you? How do you know?

What might we look for to discover the emergent quality of self? How do we look for it? Is there a golden thread of similar experiences, models and theories that makes sense and, by doing so, bring us to an advanced state of awareness?

ZERO to ONE, Making Our Way Toward a Conscious Civilization is an exploration of the systems already present within each of us as well as some others nearby. We've had little open access. According to ancient wisdom; the prudent path appears to those who are intent on finding it.

Whether I found the path for me is inconsequential, though obviously I've at least found some breadcrumbs that are leading me there. Beyond dealing with Nature's course, peace is the main course of action for us now. How do we achieve it?

It's also been said that truth is transparent for those who can perceive, and we can all perceive it. It has a frequency that causes a resonance in our perception apparatus, our gut. I encourage you to have gut checks along the way, as you no doubt will just in the course of reading.

Let the pause button be present. When it happens, let the natural question emerge, then hit the pause button. Let the answer emerge as well. Thinking too much is a problem; we over-think most things if given the chance. If what I allude to in this material is true, there is a natural send and response system that works beyond our imagination.

This system is of deep-design and has its own timing; sometimes linear, sometimes non-linear and sometimes even non-local. If those terms are unfamiliar, you'll enjoy the read as you discover your familiarity with them already, in your own terms, through the descriptions and ponderings I offer.

If this frontispiece intrigues you, I suggest you read Stubbing My T.O.E. on Purpose – A Seminal View of Consciousness, Cosmology and the Congruence of Science and Spirituality.

Enjoy your exploration!

Copyright © 2019 & 2021

Bruce 'Zen' Benefiel

Publisher: Be The Dream LLC

ISBN: 9781692812386

Contents

A Life of Coagulating Possibilities

This book started out as a chapter for the FREE second volume of Beyond UFOs. The first volume contained survey results from thousands of 'contactees' and/or 'experiencers' of the UFO phenomena up close and personal. It was several years in the compilation, including articles from experts in the field.

I am one, too, 'experiencer' that is. The request for the chapter was an opportunity to compile and report some of my experiences. The process led to an expanded view of our capacity for such things to occur based on the built-in systems we have from birth, yet with purpose unknown. Perhaps it is knowable and our systems are designed to reveal it.

Containing my flow within the guidelines for the chapter length spilled into a much larger volume and perhaps much needed expansion on the theme of holistic systems. It was cathartic and informative, as I'm sure it will be for you. Let's take a deep dive together and see where it goes.

The expanded version is a good stand-alone volume that deserves publication, whether the original chapter is included in the book or not, let alone its final publishing. In today's world content is king and repurposing, queen. She always knows just what to do and when.

I've written copious articles, blogs and books to date, the majority of them attempt to impart some experiential wisdom in a field where non-experiencer research and sensationalism has often been the norm. Ufology deserves better. I admire people of integrity, like Dr. Edgar Mitchell, whom I call friend. He was an example of being outspoken and stalwart for truth.

Though I've earned a Master of Arts in Organizational Management, a Master of Business Administration and an honorary Doctor of Divinity, among professional credentials,

I'll forever remain an eager curious researcher of the human experience and continuously seek to correlate and understand my experiences which travel beyond such limitation. Regarding engaging non-human intelligence(s), I've been active since childhood in more ways than most can imagine.

I also maintain an avid interest in how people, places, and things might work together to ensure optimal results, and am a big fan of Csikszentmihalyi. It is because of the time I've devoted to making sense of this mysterious world from its most mundane to its most incredulous. I use that knowledge in my professional life as a transformational coach, digital specialist and construction partnering facilitator. Truth is often stranger than fiction and it can be presented simply.

Including projects for large construction firms as a pre-construction team building facilitator, I've made it a lifelong mission to both observe the "unseen" and to participate in a robust array of 'contact.' It eventually paints a more detailed picture of this place that offers such little explanation. While my professional training has taught me how to recognize certain opportunities, it has been trial and error, an extensive sense of self-awareness, and an ability to keep an open mind that's convinced me I can and will always learn more as the journey never ends.

Consciousness is defined as **the state of being awake and aware of one's surroundings**. From the perspective of non-human intelligence, this expands the playing field across much of the electromagnetic spectrum. According to our current science and understanding, reality is a vibrational field of dynamic proportions that seems to respond to attention, intention and interaction. Being still is a form of interaction.

The frequency of people, places and things determine the resonance they have at any given state of awareness. I hope to explore some possibilities in this book and expand the perception of a greater reality in answering the three questions presented. This was originally a chapter for a compilation to be published by the Dr. Edgar Mitchell Foundation for Research of Extraterrestrial Encounters (FREE). I think it is valuable for the exploration of life, an expanded reality and the potential we have to evolve toward a higher order even in our lifetime.

I met Rey, the Executive Director of FREE at the International UFO Congress in 2014. I had previously spoken to him on the phone and agreed to receive and deliver some materials to the event for him. I joined him with an air of excitement when he invited me to share in his space on the vendor side of the table. I had been a book and cd (my own) vendor there as an individual and sharing a spot with others from 2010 to 2013 when we launched UfologyPRSS.com. In 2014 I was taking a break from disappointing results in breaking even.

I could relate to his desire to make disclosure a moot point by bringing together so many people who have so much to offer, especially with the results from the surveys he was having developed. I was happy for the opportunity to introduce him to a host of characters in the Ufology arena to get him known. He was an unknown at the time and I'd made friends with many over the years and the personal relationships allowed for a swifter introduction.

Rey's ability to tend to such a work 'assignment' from elsewhere while simultaneously maintaining a job, a family, along with the tireless load he carried while working on his PhD surely presented him with daunting challenges that demanded relentless efforts. I felt honored when he asked me to write a chapter for the next volume (previous was *Beyond UFOs*). I knew this compilation of perspectives might one day become the turning point in making sense common as it relates to intention, nature and purpose of interaction with non-human intelligence. It can be a pretty freaky experience.

I'm grateful for FREE's work in forwarding this effort to distill our shared experiences and allow for us to engage one another through the Contact Modalities in more efficient ways. I have little doubt that the skills required to compress, truncate and succinctly craft this lifetime of work into such a relevant presentation would be worth the time and effort invested and wait in excited anticipation to assess the congruency of our combined answers.

I ask only that you set aside your own personal knowledge and experience as well as continue to challenge your own conclusions so that you might always remain open to new ideas and differing perspectives. If this material works to corroborate your thinking or offers a correlation to your own experience in some small way, I shall consider it a great success and take a great deal of pride in the small part that I played to assist you in to raise your awareness.

I remain grateful for this experience that allows for me to not only witness such amazing work by such extraordinary people but to also have been fortunate enough to play a part of it.

It's in your hands now so let your gut become your guide and enjoy the journey. If you are still not quite sure how, there's a chapter further in that speak to qualifying information.

A Personal Experience of Major Significance

My interest in space started as a child, and in Junior High I wrote a research paper on the Gemini and Apollo missions, through Apollo 11. I followed the future landings with intense interest and, as I learned of Dr. Mitchell's Institute of Noetic Sciences, I longed for a conversation with him about my own experiences in hopes of corroboration of my learning as well as to hear about his own profound and life-altering events.

One of the things that is prevalent in my life is the opportunities to enjoy wish-come-true moments have been repetitive enough not to be random. This is one of the more obvious that wasn't at the time. I hadn't noticed the pattern. I have since and it only increased the sense of awe in life.

I met Dr. Edgar Mitchell in October of 1997. I was running the Prophets Conference in Laveen, Arizona at the time, recommended by an old friend, Jim Diletosso, as being 'the guy' needed to do the job. I put all the logistics together in three days, with delivery and set up on the fourth. It was a mostly outdoor even that drew 5,000 patrons with 30+ speakers and 120 vendors for a weekend.

I notice Edgar as he stepped into our 'green room' which was a secluded outdoor area. I watched as he pulled a pouch of Drum rolling tobacco from his back pocket and, it just so happened, I had one as well. A plan instantly appeared.

I pulled my pouch out as I walked up, asked if he minded if we shared smoke, and introduced myself and role in the event just to set him at ease. He laughed and admitted that it was extremely rare to find another who rolled their own, let alone carried around a pouch of Drum tobacco, too. It felt like an instant friendship emerged.

We talked about tobacco and its sacred use among Indians which led into indigenous ways in general and their closeness to Mother Earth and Father Sky. I wanted to let him know the Apollo missions were important to me, so I mentioned I had done a research paper on Gemini and Apollo missions in junior high to let him know I had more than a casual interest.

He started off shared his awakening moment of realization as well as his most notable political statement, "You want to grab a politician by the scruff of the neck and drag him a quarter of a million miles out and say, 'Look at that, you son of a bitch."

We went on to talk about the shared understanding of consciousness from experiences we'd had and the comfort level between us felt inviting and warm. I felt free enough to ask permission and then share some personal events that shaped my perspectives and views about consciousness and my belief that there must be a way to engage our whole being in the experience of reality.

We both agreed we are only using a small percentage of our brain/mind connection and the greater self-awareness, which brings greater awareness of All That Is, was a quest worthy of our lives and work. It was such a joy to have these conversations over the weekend, truncated as they were due to activities and responsibilities I had to manage.

I didn't think about it at the time, though one can get a sense of integrity and trustworthiness from these kinds of discussion and words one has and uses. Real experiencers and folks with in-depth understanding talk much differently and to Edgar I'm sure it was easily recognizable. I'm honest and transparent to a fault, willing to question my own perceptions with fervor. He opened up to me in ways I certainly did not expect.

He shared some private information with me and asked me not to say a word until after his death. The sense of trust in his eyes when he said that was undeniable. The details were short and sweet. Beyond the profound views and statements he'd made to date about his 'enlightening' experiences from his trip, there was something he could not speak about in public at all. I understood the notion of secrecy with NASA and why.

Edgar explained that after separating from the command module, there were silver metallic cylinders that came off the surface of the moon, spiraled around the LEM and followed them almost all the way down to the surface. This was fascinating and corroborated things I'd heard about Aldrin's and Armstrong's sightings of similar objects, though they were on the surface standing upright on a platform.

He wasn't sure how many there were because of the limited viewing from the tiny LEM window. He had a lot of questions running through his mind, from specific details of what they were to how they were controlled. He thought there was at least a few, though admitted it was probably only one and estimated the size to be about a meter in diameter and several meters long. Whether it/they were piloted or controlled from the surface he couldn't determine. He knew beyond a doubt that they were not anything from Earth.

The cylinder(s) disappeared by the time they reached the surface. The details seemed to match up with stories I'd heard about earlier missions. I believe it was Buzz Aldrin and Neil Armstrong having seen these same types of cylinders standing upright on a flat structure just beyond the demarcation of the 'dark side' of the moon. I'd heard that a decade prior and Edgar validated it, at least the similar type object.

Unfortunately, he said, all the astronauts are under agreement to remain silent. They are prohibited to say anything in public about what they saw in regard to extraterrestrial objects or craft. Those agreements can cause major damage for the astronauts if they violate them.

Now, after some rumination, I'll include some other considerations and personal stories, some of which you may have read already if you're aware of me at all. I'm assuming you are not and have not.

Laced within the original text of the chapter for FREE, I've inserted pieces I feel are relevant to the concept of making our way toward a more conscious planetary civilization.

Perhaps some hearts and minds will open to new possibilities as well. This book isn't about creating procedures and protocols for planetary management. It's about raising awareness for growing together in our ability to do so, free of constraints and full of possibility.

It is my theory that the understanding of our natural design and the ability to explore it without constraint will also invite a deeper awareness and collaborative collective mind flow in those who acquiesce to their own truth and Being, which I called The God Participle in a previous book. Its where an 'ing' is added to a verb and is used as an adjective or a noun.

I love the English language. My adoptive mother was an 8th grade English and Literature teacher for 29 years. I became one, too, by default when I took on responsibility as a permanent sub for a high school English teacher as my first teaching assignment and then as part of an entire high school curriculum for a charter school a couple years later.

But I digress…

What Are the Three Questions?

They are no doubt many questions in the hearts and minds of many who've ventured or been drawn into the study of consciousness, Ufology, or quantum physics as well. There's a bit of excitement about the capabilities of consciousness. These questions embody a theme and work that will lead us, I believe, to a new living awareness and greater capacity for harmony among people and planet. The latter, our destiny.

What is non-human intelligence?

What is the relationship between Consciousness, our Cosmology and Contact with Non-human intelligence via the Contact Modalities?

Is there a possible unification theory of the 'paranormal' contact modalities? Are all the contact modalities interrelated?

What is unique for me in answering these questions is the book I recently fine-tuned and republished under the title, Stubbing My T.O.E. on Purpose – A Seminal View of Consciousness, Cosmology and the Congruence of Science and Spirituality. It was also part of a 28-year wish-come-true.

It took me over a decade to write it, first publishing it in 2010 as *Zendor the Contrarian* with the same subtitle. My wife, Luba (wed in 2017), and I are huge fans of Tom Campbell's work and I wanted a title to reflect a little more intelligence, humor and multiple meaning beyond a double entendre.

I explore much of what the questions are about in my book, viewing them from a variety of perspectives with critical questions and potential answers. It is a pseudo-autobiography that took me in a lot of directions and gave me a plethora of cathartic experiences of the time it took to write it.

Because of the myriad of styles and types of experiences I've had, my intention was to peer into these topics with an open mind and nimble fingers, share some thoughts and maybe garner resonance. Life *is* an experiment, right? Let's have some fun together, shall we?

I've kept a healthy sense of humor and pretty much anything is fair game to make fun of in appropriate and sometimes sarcastic ways. I learned long ago a paradox of truth - everything is sacred and nothing is sacred. Our experience is either desirable or undesirable and we usually make our choices toward the desirable, toward the resonant.

If we cannot make fun of ourselves then, imho, we're not open enough to explore the possibilities that life offers us to explore more fully. Living is a visceral experience, after interpreting those sensations we ask questions and make choices. An insatiable curiosity is best for making better choices through asking better questions and really being present for the answers to flow.

In exploring these questions, we'll look at a plethora of perspectives from the subjective to widely accepted schools of thought regarding physiological and neurological systems in the body. These systems, combined with a reasonable paradigm shift in awareness can accelerate the development of self-awareness and one's ability to negotiate the often arduous requirements of life and work.

I use my own life as a barometer, an experiment of sensing, and answer questions or illustrate understanding. Using current knowledge and understanding, I observe and report on life experiences that imbue the response with humor, insight and perhaps a bit of wisdom. I often leave questions unanswered, too. It suits an insatiable curiosity. I trust you'll enjoy this journey.

What is Non-human Intelligence?

First of all, the current state of affairs in our world might indicate we aren't as intelligent as we think. Central to that intelligence is making sense, which many of our current systems do not, even with the increasing IQ scores. We have a dichotomy of belief systems and the cognitive dissonance in our society that proves we are challenged to follow any prudent path toward harmony.

On one side, a fear-based system about the command and control of resources lead by a small but powerful number of corporations and people, including our media. On the other side, a love-based system about learning how to create harmony among people and planet that fulfills an evolutionary path consistent with advancing intelligence that includes thousands, perhaps millions of people globally.

The former has little intelligence beyond understanding how to manipulate people, places and things. Manipulation doesn't always have detrimental results. The latter operations take us to a new living awareness that engages an understanding of all things being connected, not necessarily knowing exactly how. It leads to studying systems and learning how they integrate.

We've got the technology now that can share that information immediately and reach millions of people through social media networks. Perhaps our next step is a peer-to-peer collaborative information system able to store and serve up challenges and solutions in perfect timing for the next phase of our global village to evolve. To be able to position ourselves to look from this perspective I believe is imperative in order for us to have a sustainable future for us all.

Non-human intelligence would seem to include the 'kingdoms' of order, from the elemental to the universal. Non-human intelligent beings seem to have a holistic understanding of systems beyond anything humans can imagine. In the development of understanding, intellectual or spiritual paths, it becomes apparent that humans aren't as high on the coexisting scale as we'd like to think. We seem to have misunderstood the 'have dominion' for 'have control' instead of 'have good stewardship.' Only recently has there been a 'good steward' ideal that is being presented across multiple frameworks.

It's been our habit to condemn and/or destroy what we don't understand or have never encountered. We've also allowed flora and fauna to become extinct as well as decimating habitats and ecosystems. It seems counter-intuitive to 'intelligence' and yet the pattern is and has been ubiquitous for millennia.

Is it possible that non-human intelligence, in whatever form, might be intimately connected and humans, as 'free will' characters in the play, have just been afraid to explore that possibility without fear and with an open mind and heart?

A recent interview that Jaime Maussan did with Dr. Michio Kaku at the World UFO Congress in Barcelona, Spain revealed a new perspective on 'disclosure' and the burden of proof. Military videos released show that UFOs are indeed a reality. Now the 'burden of proof' is on the military and/or government to prove they are not real. Unfortunately, that process leads to a consideration of threat.

Dr. Kaku also goes into the Kardashev scale and notes that advanced civilizations would have technology that we probably would not be able to understand, let alone recognize, and/or communicate with them in any kind of human ways we engage.

I had also written a blog post on frequency a couple of weeks prior to the event, noting the probability that Type III civilizations would more than likely use vastly different types of communication methods that were effectively 'dumbed down' in order for humans to be able to recognize the attempts.

I believe strongly that several methods are obvious, from the crop circles to the high frequency tones that most experiences report accompany their 'contact' events.

In an interview from August of 2018 for *New Thinking Allowed* with Dr. Jeffrey Mishlove, titled **A New World Order**, we explored the notion that these frequencies hold vast amounts of information. An experienced meditator, for instance, is able to find repose in the silence within. In this same place there is a frequency that most are able to recognize, or at least acknowledge it is there in the background, underneath the visual or auditory display encountered during deep meditation.

It has been my limited experience that this *frequency* can be focused on and, as one does, another world opens on what we've called the 'inner planes' of consciousness. The experiences are non-linear and non-local, yet are often called 'downloads' because of the rapid-fire information exchange that occurs. Regardless of what the process is called, the events happen and people are not necessarily equipped to articulate the experience in any observational way.

Non-human intelligence across the gamut of types of civilizations and the many worlds we are told exist from the ancient texts and modern explorers has traditionally been kept and labeled in categories that severely limit its exploration, let alone open discussion without polarity paradigms. Perhaps as you read on this stigma and your own resistance to explore it in greater depth with dissipate or even dissolve.

It really is a fascinating world to explore. What is most challenging is stepping away from the traditional views of human intelligence and how it works. We think and act in linear fashion, with rare exceptions. Non-human intelligence seems to behave in non-linear and non-local ways which confound the mind because it has no way to process it.

However, with practices like meditation, this process can be at least be understood for how it works and the observer within begins to adjust to the new environment. I've often wondered if there is an inner world or worlds that are like our outer world(s).

Sometimes I see the same with eyes open and closed, too. Is that liminal space beginning to be more exposed?

Is there some evidence we might be able to explore, some communication from 'elsewhere' that has significance? It appears there is at least one exploration from the 1950s with a scientist that was in charge of Canada's UFO investigation program. The program was funded by the Ministry of Transportation and regarded as a leading-edge work.

His name was Wilbert Smith and he kept a diary of his encounters and the in-depth conversations (telepathic) that happened. His memoirs were published two years after his death in 1962, and are called 'The New Science.' (https://bit.ly/WilbertSmith)

The Information Age

"Information is notoriously a polymorphic phenomenon and a polysemantic concept so, as an explicandum, it can be associated with several explanations, depending on the level of abstraction adopted and the cluster of requirements and desiderata orientating a theory. Information remains an elusive concept."

https://plato.stanford.edu/entries/information-semantic/

Information feeds the growth of intelligence. How much information do we really have about our human form, fit and function in this reality? Wouldn't it make sense for advanced races, non-human intelligence, to perhaps have more information as a result of their prior evolution long before us?

Just consider the thousands and perhaps millions of years they are ahead of us. It gives us pause for a moment to consider our tenure in the universe, looking at our current information curve and state of affairs, noticing a similar curve of awareness leading up to 2012 and recognizing that there are civilizations thousands, perhaps millions, of years ahead of us.

How might we understand non-human intelligence? Are we trying to mix apples and oranges? Perhaps it's just a riddle and we forget the option of juicing.

Throughout the 'Contact Modalities' we have examples of how non-human intelligence has interacted with people from all cultures, religions and walks of life for quite some time. Non-human doesn't mean exclusively ET, either.

Most of the time we've chosen to categorize and separate, a very human-centric left-brained trait, instead of looking at them as a holistic system. The very idea of everything being connected, which many know to be true, is muddled by the reticence to expand the view, most likely because humans feel like they have to have answers in hand before approaching a subject. Curiosity beyond cursory is missing.

What if we didn't have the answers? What if we didn't even have the right questions? By all accounts, with rare exceptions, this non-human intelligence seems to operate above and beyond the notion of any human control or understanding. What, then, can we do?

Heck, we're suspended in space on a planet (perhaps with its own consciousness) hurtling through space at breakneck speeds circling a star that gives us life.

This same solar system is also in a far arm of a galaxy, moving around its center all the while moving around the universe with other galaxies.

All of this is happening with relative harmony; rarely does anything collide. Think about the magnificence of that reality for just a moment and let it sink in.

If we look at many of the modern scientists there is a consistent notion that consciousness, that which occupies the space between the proton, neutron and electron on a very basic level, manages the whole show. We don't know exactly how humans fit in, yet the mathematical precision becomes evident. We tend to reverse-engineer everything, destructively and without regard in most cases.

We're great at taking things apart and developing equations that explain how things are put together. For instance, did you know that the equation that was discovered to reveal the creation of the carbon atom has 32 decimal places in its variable, the number that actually makes the equation work?

What if there was that kind of precision within a different state of consciousness and, over time, this same consciousness is not only understood, it is accessible for interaction. That's what we're exploring now.

Later in this material is a chapter on Frequency and how it might relate to the contact modalities. At least for the moment, let's just consider what we already know and expand on it: everything is vibration. This means it has a frequency.

There is a unified field of these frequencies or vibrations that is naturally occurring. Could there also be resident frequencies that, when we ascend to new levels of consciousness, actually trigger genetic material releases in our DNA?

Valentina Mironova, a Russian Academian, gave a dissertatoin recently that reveals her discoveries of the last decade or so that offer quite astounding evidence that there is a genetic mutation happening in humanity, an ascension of consciousness that includes co-creation processes. We can actually interact with others and the physical world by focusing our attention, intention and interactions.

Re: https://youtu.be/sMxvNJMiJHs

The Human Age

Just for a moment, suspend your human thinking, the notion of needing to explain everything from the perspective of what you already know. For most of us that is *nearly* impossible, emphasis on the 'nearly.' That is what usually happens when we encounter something outside our experience.

We want to leap into explaining it somehow, often bereft of the actual facts. We lack the knowledge or understanding to do so because of our subjectivity. It's just hard to observe and report while in an experience, though there are ways. William Swygard developed a process to do so in the 1950s as well.

We're used to a physical reality with certain rules we can count on. What of the moments in nature when we sense a synchronicity with our activity or, even better, our thinking? What about, in a moment of pure innocence, a question we've never thought of comes up and, mysteriously, an answer immediately follows? The latter may involve a 'sense' or a 'voice' or even imagery; most often it comes with a combination of those, right? Is that *our* thinking?

In modern times most of us think of non-human intelligence as types of beings we occasionally encounter and aren't quite sure how to categorize. They could be angelic, demonic, extraterrestrial, and multidimensional or even something else entirely. We don't really know, yet we do have encounters with it/them in a variety of ways historically.

Some apparently are prepared before entry into this world, or have a genetic predisposition, that they are engaged from a very early age. As much as we know we still do not have the answers. How the heck do you test for 'alien' DNA anyway? How would we know what it is? What of the depth of sensory ability we have barely come to understand; referencing the 'clair-alls' (clairvoyance et al) of perception?

Evolution promotes a lot of things, over long periods of time, and we as humans tend to narrow the scope and timeframe to those we are familiar with on Earth. The last century or so has given us an expanded viewpoint that takes those thousands, or hundreds of thousands, of years on Earth and makes them pale in comparison to the millions of years other civilizations have had to develop elsewhere.

Perhaps they are closer to our galactic center as well since our universe is known to be expanding? How can we fathom light speed travel or even the speed of thought when we've barely reached past supersonic in our own world?

We tend to apply the science and technology that got us this far in trying to understand the technology for space/time travel. Even that is pale by comparison when we consider the technology of our own bodies, our minds and how we interact with the world(s). Tom Campbell speaks of his efforts to explain right-brained thinking to left-brained listeners in his talks about his books.

I wonder if the corpus collasum, the membrane between the two hemispheres, might have more capability of bridging the hemispheres than we understand yet. It certainly is a conduit for neurotransmitters. Do we have a choice in what they transmit? Can we? How? Bruce Lipton's work is noted.

I'm reminded of the admonition that Einstein gave us in dealing with new things – 'We cannot solve problems with the same thinking that created them.' I believe that applies to math, science and technology as well as the human anatomy and neurophysiology. Dr. Evin Laszlo speaks of the 'Higher Self' in his explorations across nearly 200 books.

In the case of non-human intelligence, we attempt to understand it through conventional human-centric models that just do not apply to 'their' worlds. We do have free-thinkers in various fields that explore other options and ask questions beyond the nuts and bolts scenarios. Their numbers and voices are growing as the conversations deepen and questions beget questions.

The first volume from FREE, *Beyond UFOs*, is a perfect example of expanding the vision of what we know about consciousness and the nature of UFO encounters. Subjectivity became data for the first time, perhaps. For years we've been caught up in the study of the phenomena, yet the focus has been on physical evidence and not on the pilots or occupants of the craft. Most have been afraid to speak up.

Now with the FREE surveys, it is even further defined by how these occupants have interacted with humans. We're moving beyond the phenomena of sightings to the curiosity toward the lifeforms in these UFOs. Who are they? Why are they here? What do they want? How can we engage them and get to know them better? Do we need to fear them?

In the same fashion, the study of NDEs has been from a clinical perspective with the parapsychological implications a lesser priority. I was the first to introduce the extraterrestrial element into The International Association for Near-Death Studies at their 2010 conference in Denver, Colorado. I had a great conversation with PMH Atwater as a result.

There is an opportunity to get a little off-center after NDEs, even though the experience brings great knowledge. It's the wisdom that is often a little lacking for some time, hence, the Messy-Antic Complex. The presentation is on YouTube, just search for 'iands zen benefiel,' or, if you've got the e-book version, just go here.

Apparitions of the deceased tend to cause more religious concerns and discussion about the event with little exploration into the reasons. That is not to say reasons are not explored, just not as regularly. The event usually trumps the evaluation as far as stories go. We tend to get caught up in the excitement and are bereft of the capacity to observe and report on the events through open inquiry. We just can't be still enough to find our center and report, let alone converse.

Perhaps that is a pattern in our behavior toward the unknown and across the myriad contact modalities; we're lazy in exploring further and really understanding how they function in our lives. Obviously there is a growing interest as we face our mortality in the idea of global warming and its implications. What change is imminent?

What if a greater functionality also created solutions where none have been found? It certainly gives us pause for consideration. I'll digress for a bit to fill in potential gaps proactively and offer a more complete view.

Phases of the Moon Child
Phase One - Experience and Inquiry

I had a unique opportunity as a kid because of circumstances, to explore non- human intelligence through investigating my own life and origins. I was orphaned at birth and adopted into a great family about six weeks later. A few months after my fourth birthday, my adopted sister arrived.

My parents thought it wise to explain the process of adoption, as best they could to a child. Even then I was ruthless with questions of where I came from, not only biological origins but from 'heaven' as well. I had these conversations inside mostly, as I knew it would disturb my parents, somehow.

I had conversations with 'God' sprinkled with questions about my origins and who my mother and father in heaven were. This curiosity set the stage for my life, perhaps a combination of destiny and pre-disposition. Life got really weird.

One night a few months after I was told of my adoption, I was standing on the landing of our stairway, peering out the small square window that looked out over our front porch, with my elbows on the sill and my chin resting on my clasped hands.

"Hey You!" came a voice so loud I just knew Mom, sitting a few yards away, just had to have heard it. It was loud and low-pitched, resonating in my head and not spooky at all. It had a sense of authority that was undeniable. I wanted proof.

In an instant, and with no further engagement, I spun around in my excitement and asked my mother if she heard it. It was so loud I knew she just had to have heard it. Nothing... but she didn't deny it. She explained it away as probably being a 'peeping tom' yet didn't react suspiciously at all. It was one of those parental shrug-offs that children often get when they've discovered something and just want to share.

Thinking back about the event, as I have numerous times, it is surprising to have had such lucid thinking about it. I knew I heard the voice and no one was going to convince me otherwise, yet my response to Mom wasn't to try and convince her. I remained quiet and began asking questions in my head; lots of questions.

Someone answered always. I surely was not going to tell Mom about it. So I began to develop a relationship with it, conversations only taking place when I was confused about something and would ask questions. I grew up thinking this to be a natural thing and that everyone did it.

I think that introduction to non-human intelligence (a discarnate voice) helped to open my mind to other things. I knew there were things that happened beyond the physical world and that most people never talked about them. I never did either, at least until my first year in college.

From my childhood, spontaneous and sporadic events throughout those years included 'psychic' experiences across the spectrum of the 'clair-alls' as well as a two-year period of recurring trips from my bed to an orange cigar-shaped cloud at night. They started just after my 8th birthday.

Everything culminated it seemed, on a Tuesday afternoon, November 11, 1975.

Into the Light and Beyond

I had come back to college (Ball State University) after fall quarter break. During my break I went to ask my high school sweetheart to marry me. I'd broken up with her just before starting college, realized my foolishness and hoped she'd understand and accept my apology for such behavior.

When I reached her home I found out she was already married. I was heartbroken. I returned to school and after much misery, hit my knees in prayer a week later. Empty, I asked to know 'Truth' and was willing to die for it if necessary. I meant it with every fabric of my being; heart, mind, body, spirit and soul.

It seemed like it was about a week after my prayer, on that Tuesday afternoon, that I returned from school to the honors dorm where I lived and proceeded to perform the afternoon ritual. Some of you may get a kick out of this one; a bong hit and album side meditation. Hey, it was the 70s.

That day it was Journey's self-titled album. During 'In the Morning Day' (second song), right when the lyrics ended, that same voice I'd known since childhood asked, "Bruce, are you willing to die for what you believe in?" I locked up for a moment, and then realized I'd better answer the question.

My first thought was 'Christ Consciousness,' but something inside me still felt a bit empty. Apparently, I thought there must be a 'fullness' to 'Truth' that I didn't feel it, yet. I dismissed that thought and moved on to 'Cosmic Consciousness' which seemed to feel expansive, even more than 'full,' and I said, "Yes!"

Music seems to facilitate a lot of experiences for people. In that moment there was a guitar riff that sounded like a supersonic jet going by or a rocket ship taking off. I felt a slight tug from my abdomen and found myself looking at my body draped across my dorm room bed just as I was doing at the time. I could still feel the movement away from my body, so I turned to look where I was going just to keep my bearings.

Before I could even turn completely around, I was engulfed by 'White Light' and instantly felt like I was 'HOME.' It felt like no other feeling I've ever felt in this life. It had a 'sense' of iridescence and effervescence with a high-pitch that was oh, so amazing to feel. I've had that sensation a few times since; embracing my soon-to-be wife in a dream was close, replete with a sense of Oneness, though it was decades later.

It was like a complete union with everything I'd ever imagined as God or even my celestial family. There is a certain sense we have with parents that is deeply connected and trustable. I felt like I expanded to infinity with it... and I wasn't 'dead,' though by most NDE accounts I was.

I could still think and was aware I could think. I recognized the self-awareness and so I knew I wasn't 'dead.' Still, there had to be more than just a blinding white light, right? So I asked, "Is there more?" It was more like a question to self than anyone specific at the time.

Here's where it got really weird. I felt a slight rush of energy, a movement, and found myself in a nearly opposite environment, surrounded by points of light on an indigo background. As soon as began to inquire of what they were I know they were points of consciousness, whether in body or not I wasn't sure because I knew I was not. I had the sense they were, somehow and somewhere.

At that point the voice resumed and instilled a deep sense of awe in me away with this delivery:

"These are those you are to work with in order to facilitate a new world order. It will happen in your lifetime. Know this to

be true. Your path will be full of trials and tribulations. Have faith and trust that everything you need will be there at its appointed time. Trust and allow."

Almost instantly at the end of the statement I felt another rush of energy, this one much stronger than the first. I was back in my body taking a gulp of air. I kept my eyes closed as I felt a kind of newness and re-integration with my body simultaneously, descending back into the physical world.

I lay there, eyes closed, for the rest of the album side just feeling this amazingly vibrant sensation coming from or focused in the center of my body. When I opened my eyes my first thought was, 'What the f... just happened? How am I to do this? What do I need to do to prepare?'

The first stop, after telling my parents, was to see a psychiatrist. I wasn't out of my mind or acting crazy, just trying to understand what had happened and need to share the experience as clearly as possible. It was too much for my parents, and their concerns went to drug use, I'm sure. I had been experimenting, as many did during those times.

After some serious questions and listening to me for several weekly visits, Dr. Abel eventually tells me I'm not crazy and that I'd experienced a 'spiritual awakening.' I was 18. He explained that most people do not have them until their mid-40s, if they ever do, but rarely does one have them so young.

His best advice was, 'don't talk about it,' warning that others just would not understand. Interesting twist that only recently came about as my half-sister found me through a DNA match in early 2019. She had seen the same psychiatrist several years before me. Finding my birth parents was another trip.

I think Dr. Abel was a blessing for both of us and, for me, more so. He was so right. Over the years the perception most folks had was that I thought I was some kind of savior or insane, or both. The opposite was true for me, thinking that 'facilitate' was a lot of work that held no concept of being anything but a worker bee with a mission; serving the queen. In management circles it's called being a Servant Leader.

It is not a hive mentality, though, as each of us has a particular skillset designed to empower the evolution of our form, fit and function in reality, together. In my opinion, this is the crux of the intelligence we call non-human, inviting us to think outside our box and embrace a universal aspect we've yet to consider fully. Obviously there is a natural order to things, right?

We are all in a relation-ship seeking safe harbor on the ocean of emotion in our experiences. For me, that tumultuous journey was exemplified the following year in college. I fell into a near complete disregard for the 'practical' things. I dove into paranormal experimentation with others that 'appeared' in my life.

After a serious of fortunate/unfortunate events, primarily purchasing two sets of drums with my room and board money which required me to move out of the honors dorm during winter break, I was nearly destitute. I'd found a house needing a caretaker for free rent and moved everything without my parents knowing. I did have transportation and a job at the honors cafeteria, so I had gas money and little more for food, though I didn't eat much at the time.

I lost my transportation a week later on the way to school with friends; being ran off the road into a telephone pole trying to miss an oncoming car that had veered into my lane. An off-duty cop was driving it, having accidently swerved into my path. I clipped off a rotting telephone pole that has a brand new one right next to it. Fortunately, our momentum wasn't stopped by the new pole, which could have been catastrophic as none of us wore seatbelts.

I got rides for the rest of the week from others, until a major snow storm and high winds with a 77 below zero wind-chill factor hit Muncie. I hitchhiked into school in white-out conditions, avoiding death by exposure, and that night was accosted at a frat house in my attempt to stay warm.

I was taken to the emergency room for stiches, being 'saved' from a major beat down by the campus police that showed up just in the nick of time as my accoster was readying a blow while I was on my back in the snow. The timing was impeccable and evidence of a 'protection' from an unseen hand, perhaps. It was a bit spooky cool.

Dad had informed a family friend, then the university's Administrator, of my 'condition' and campus police were told to contact him 'if/when' my name ever came up. He was in the waiting area when I came out from getting stitched up; NOT a happy camper at 3 am. I certainly was appreciative.

He told me Dad was on his way and a half-hour of grueling silence later, he arrived. Obviously concerned, Dad let me know I was not going home yet. I figured it might be because of a possible concussion and I'd have over-night observation. I had no idea I'd spend some time in the psyche ward.

Explaining my process, although I understood it perfectly, sounded to the hospital psychiatrist that I'd lost touch with reality. Oddly enough, he was also a family friend and had a cabin next to the one we rented on Lake Tippecanoe in the summers. I didn't remember him immediately and that might have raised his suspicions about my state of mind.

After an arduous process and finally convincing him of my sanity, which meant shutting up and telling him what he wanted to hear, I was released six weeks later. They put me on Thorazine (500ml x 4 a day) for three weeks, then when I had the 'miracle' breakthrough the dosage was gradually decreased, though it took months to complete the process.

Believe it or not, even on that amount of Thorazine, I was still able to beat the male nurses at ping pong. I should have been a lump in the corner. After my collegiate days and recovering from the disappointing experience with modern psychiatry, I still knew that what I'd experienced was real. No one was going to take that from me.

What it did leave me with was an inability to be still for a while. I'd get so anxious in talking to people that my body would tremble. I was afraid of going back. It was horrible, let alone embarrassing. I'd been gregarious and unafraid to engage prior, a real extrovert.

Over time, though, I got involved with the Jaycees, which gave me speaking opportunities, and starting playing drums again as I still had them, fortunately. I joined a band and had gigs so my anxiety gradually disappeared as I loved to play and talk about things that mattered beyond my own story.

Phase Two – Married with Children

A couple years later I was married with a one year old daughter and another daughter on the way when a decision point came. The auto industry downturn decimated our local economy, leaving few jobs and businesses failing. I was working as a machinist, a meat-cutter and playing drums in a Christian rock band.

In late March of 1981, all three jobs fell through in the same number of days. What was I going to do now? After the last was gone, I got up the next morning and went out on our front porch, threw up my arms to the Sun and exclaimed, "Okay, I'm listening! Where do you want me to go?" I had no idea and was shocked when I heard an unmistakable voice, "Phoenix." I got my answer.

We had some friends, two married couples that had moved there the year prior and, with a couple of phone calls, had a place to stay and job ready upon arrival. We lived with them for a few weeks until we found an apartment. A few months afterward a childhood friend named Steve, whom we all knew well, died in a motorcycle accident.

We invited the other couple that lived in Phoenix over for dinner; his ex-girlfriend, Julie, who had their daughter and was now married to our mutual friend, Bob. They had lived just across the field from us in rural Indiana, a few miles outside of town. We'd enjoyed many a bonfire and roasted sweet corn there.

We spoke of Steve's death briefly and Bob related an event that happened to his best friend from high school that had died from a brain tumor. His name was also Steve. They had made a pact that Steve would come back and let him know what the other side was like after his passing.

He wasn't prepared for it. Steve showed up in a dream one night. Bob began talking to him in the dream, realized he was dead, and asked him what he was doing in his dream because he was dead. Steve replied that he just got tired of listening to the crickets and wanted to reach out if possible.

Bob sat straight up in bed, noticing immediately that the room look exactly like Steve had it. It was now his and his wife's bedroom and had looked quite different until that moment. He woke Julie, noticed the room was normal again, picked up Charity, and went to his parent's home in town, freaked out about the whole thing. By the time he was finished I was smiling like a Cheshire cat, knowing that it was a real event and evidenced by Bob's visual experience of the bedroom for a moment after awakening.

I didn't say anything, but wondered if I could speak to the Steve, our mutual friend, and have a conversation of some kind. We talked a little while longer and noticed our daughter, Krystal, was in Charity's arms, fast asleep, so we saw them off and went to bed ourselves. It seemed my wife fell asleep quickly, so I was alone in my thoughts.

I couldn't sleep thinking about the possibility of conversation, waiting for my wife to fall asleep. I knew she would not be able to handle anything like that because of a prior event that frightened her. One of the sets of drums belonged to an uncle of another childhood friend, who'd passed away decades before (the drums still had calfskin heads).

I asked him to help me learn how to play and he responded. In the country house he made his presence known while I was at work one day, a butcher in town, and my wife called me crying and screaming for me to come home. We only had one vehicle and it was the middle of winter so she was stuck in the house with our newborn daughter of three weeks. The drums startled her awake just after a feeding, which frightened my wife even more. I thought it was cool, but that didn't matter.

Once I thought she was asleep I prayed, "Jesus, if it be your will, your will be done. Can I talk to Steve?" I started calling out Steve's name telepathically with no response, even with 'louder' calls. Then I realized there are many Steves and perhaps I just needed something more specific to get his attention somehow.

'Blab' was his nickname, so I thought I'd try it. "Blab, are you out there? Can you hear me?" "Bruce, is that you?" was his instant reply. At that same instant my wife raised up from her pillow. I could tell she moved although the room was dark. "Bruce what are you doing?" "Nothing, why?" "I just saw Steve's face and I tried blinking my eyes and he wouldn't go away." All I could do was sob with deep appreciation and relief that it was real or at least witnessed, tears of awe and joy streamed down my face.

When I regained my composure I explained to her that it has been proved medically that the body loses 21 grams of weight immediately upon cessation of life. Weight had mass and mass, form. I wondered if we can see him with our eyes open. She was having no part in that and told me it scared her and she was going back to bed; I could do whatever I wanted.

I laid back down and asked Steve if I could see him. In a moment a mist formed at the end of the bed, like a glowing grayish-white substance. I watched with amazement as it condensed into his physical form standing at the end of the bed. I wasn't sure whether to believe my eyes or not.

Excuse the language here please. "Far fucking out!" is all I could say, then "Tell me can you travel with a thought?" was the first question I asked without hesitation. Instantly he left in what appeared like 'light trails' going out of the room and within seconds returned, again standing at the foot of the bed.

The same phrase went through my head to him. Then, "Did you go back home (Alexandria, Indiana)?" I asked. "Yes," he replied immediately. My mind was at near capacity for dealing with such an out of this world event. I wasn't afraid at all. Quite the opposite; I was ecstatic! Perhaps too much so.

Gaining a little composure I thought, "Okay, I know how powerful my mind can be. Am I projecting your image from my imagination or are you really there?" I wasn't sure what would happen next, though I had to ask the question.

Instantly I felt the end of the bed move and as I focused my vision, he'd put his foot on the bed with his elbow on his knee and chin in his hand smiling with that same Cheshire cat grin I had earlier. The focus of my vision got even more intense.

I heard, "How's that?" WTF? The energy was too much at that point and I couldn't lay still any longer. I got up, put some clothes on and took a walk, continuing to talk with him while I walked a couple of miles in the middle of the night. Most of the conversation was him asking question of me about how to navigate in his new world. He didn't know he could travel until I asked him. There was more, too.

I didn't say a word more to my wife. My experiences and penchant for exploring became an insurmountable problem in our marriage. After some years and while raising a family, I was less vocal and more introspective. Outwardly I was building a professional future, doing my best to be a good father and husband, as well as performing my duties in our church. Even with all the outer effort, I was empty inside.

I could see how my attention to the synchronicities of the events during that period facilitated what many would call 'phenomena,' yet seemed completely natural in a state of non-attachment to outcome. I never knew how or where things would end up, only that I had a deep sense of trust that everything would work out fine.

I had been working as a production control coordinator for a multi-national aerospace company, in charge of $7 million in monthly shipments of spare parts, some 800 plus. I'd come off the shop floor, previously a machinist, to explore being an expeditor at the suggestion of my counterpart on the Emergency Response Team, a Lieutenant on day shift.

A few months later I was promoted to the coordinator position, leaping over all 8 expeditors on day shift, with weekly hours ranging from 55 normally to 75 at month's end. I loved it. I had pretty much complete freedom to interact with any person in the hierarchical structure of the company, with no actual authority, in order to perform my job. My performance stats were the top in the department for two years running, out of 35 people.

I learned a lot about inspiring people, how to be authentic and produce results. I went back to school to finish my Bachelor of Science in Business Administration, with two nights a week dedicated to class and a student team. We had joined the Mormon Church and I'd been *called* to participate some evenings, teach an adult Sunday school class and other duties.

I also attended a *Urantia Book* study group in my neighborhood, just across the alley behind our house in fact. What a coincidence, eh? I first encountered the *Urantia Book* in college, a deeply interwoven encyclopedia-type planetary codex and explanation of cosmic hierarchy way beyond anything available. It was a tough read, too.

My wife was a stay-at-home mom with our four children, by then, though she managed not to stay at home, finding plenty to do and not always respectful of our relationship. I can only assume that after suffering a head injury from a car accident, the brain shake broke some established neural networks. As much as I wanted, I couldn't change that. Affairs ensued and I was heart-broken with the betrayal I felt. She'd even covered her bases with our Bishop, accusing me of having an affair.

I could only be in charge of me, alas, and I had a hard time letting go of what I thought was going to be a life-time relationship as well as being a servant leader in church. Choices were made and the deep sense of love, in spite of the sense of betrayal, didn't diminish the feelings; they just didn't allow our relationship to last. I still love her, though she'd never understand that love, even today.

The only thing I could do was trust in the process of aligning with my best self and the life fulfilled by doing so. I could look in the mirror with humility, knowing that I'd given my best. The point of security and service in the exploration of self is that the unique individuation and compatibility with the collective that brings us joy and fulfillment.

One of the passages that offers some clarity and understanding from a rather indescribably delicious standpoint follows.

Excerpt from the Urantia Book:

The enlightened worlds all recognize and worship the Universal Father, the eternal maker and infinite upholder of all creation. The will creatures of universe upon universe have embarked upon the long, long Paradise journey, the fascinating struggle of the eternal adventure of attaining God the Father.

The transcendent good of the children of time is to find the eternal God, to comprehend the divine nature, to recognize the Universal Father, God-knowing creatures as they are in their spheres, like him as his is in his Paradise perfection of personality and in his universal sphere of righteous supremacy.

From the Universal Father who inhabits eternity there has gone forth the supreme mandate, "Be You perfect, eve as I am perfect." In love and mercy the messengers of Paradise have carried this divine exhortations down through the ages and out through the universes, even to such lowly animal-origin creatures as the human races of Earth.

This magnificent and universal injunction to strive for the attainment of the perfection of divinity in the first duty, and should be the highest ambition, of all the struggling creature creation of the God of perfection. This possibility of attainment of divine perfection is the final and certain destiny of all man's eternal spiritual progress.

Earth mortals can hardly hope to be perfect in the infinite sense, but it is entirely possible for human beings, starting out as they do on this planet, to attain the supernal and divine goal which the infinite God has set for mortal man; and when they do achieve the destiny, they will, in all that pertains to self-realization and mind attainment, be just as replete in their sphere of divine perfection as God himself is in the sphere of infinity and eternity. Such perfection may not be universal in the material sense, unlimited in intellectual grasp, or final in spiritual experience, but it is final and complete in all finite aspects of divinity of will, perfection of personality motivation, and God-consciousness.

This is the true meaning of the divine command, "Be ye perfect, even as I am perfect, " which ever urges mortal man onward and beckons him inward in that long and fascinating struggle for the attainment of higher and higher levels of spiritual values and true universe meanings. This sublime search for the God of universes is the supreme adventure of the inhabitants of all the world of time and space.

Ten skills for the postnormal era

1. Boundless Curiosity — *The most creative people are insatiably curious. They want to know what works and why.*
2. Freestyling -- *We have to learn to dance with the robots, not to run away. However, we still need to make sure that AI is limited enough that it will still be dance-withable, and not not-runnable-away-from.*
3. Emergent Leadership -- *Emergent leadership: the ability to steer things in the right direction without the authority to do so, through social competence.*
4. Constructive Uncertainty — *The idea of constructive uncertainty is not predicated on eliminating our biases: they are as built into our minds as deeply as language and lust.*
5. Complex Ethics -- *All thinking touches on our sense of morality and justice. Knowledge is justified belief, so our perspective of the world and our place in it is rooted in our ethical system, whether examined or not.*
6. Deep Generalists — *Deep generalists can ferret out the connections that build the complexity into complex systems, and grasp their interplay.*
7. Design Logic -- *It's not only about imagining things we desire, but also undesirable things—cautionary tales that highlight what might happen if we carelessly introduce new technologies into society.*
8. Postnormal Creativity — *We should expect that in postnormal times creativity will have a few surprises in store for us.*
9. Posterity, not History, nor the Future — *While we need to learn from history, we must not be constrained by it, especially in a time where much of what is going on is unprecedented.*
10. Sensemaking — *Skills that help us create unique insights critical to decision making.*

Phase Three – Dark Night to Bright Light

Now we'll begin to go deeper into a reality that may cause you to seriously consider my sanity. I'd offer again just to keep an open mind and consider the possibility of things beyond our imagination and understanding that may become available to those willing to explore the possibilities.

I was being drawn back into the metaphysical world toward the end of our marriage, and not in a minor way either. I was instrumental in organizing and promoting a large conference called 'A Metaphysical Coming Together' in Phoenix at the Biltmore Hotel. I was interviewed on the largest AM station in the Southwest and got crucified by the callers for the most part, although I was able to handle the verbal assault with poise and tact. The host was impressed.

Around that time, I met the publisher of a newspaper that had state-wide distribution and wrote an article that was published in August of 1988. Right below the article was a display ad for April and Steven White who 'channeled' Ashtar and Athena, which seemed to call out to me for some reason so I went. I was curious of why I felt the urge, too.

I didn't know a thing about the Ashtar Command or Galactic Federation, let alone Ashtar and Athena, at the time. When I arrived at the event I sat next to the wall a several rows back, just wanting to listen, observe and maybe figure out why I felt like I needed to be there in the first place. There were about 150 people in the room from a variety of socio-economic levels, it appeared. The room felt inviting and safe.

Long story short, during the channeling both April and Steven seemed to 'star me down' for most of their sessions, both with eyes closed, and it made me very uncomfortable. Each began with their eyes closed and, after reaching that state of un-comfortability, I'd ask mentally if they were really looking at me or was it my imagination. Instantaneously, their eyes would open and, whether they were conscious of it or not, looked directly into mine. That was a pretty freaky feeling I must admit. How would you feel being stared down like that?

Something inside me did understand everything they were saying about the hierarchy of universe administration and structure of relationships to manage the cosmic community. I was further confronted when my question resulted in opening their eyes to look directly into mine. With all the synchronicities, the timing of internal questions and external events, I was seriously questioning my sanity.

I went up to them after the gathering had dissipated enough that no one was waiting to see them. I thanked them for the experience and asked them if they remember anything from it at all. They did not. Of course the response was not affirmative and I was left with even more questions.

My association with the Mormon Church, the New Age Alliance (membership group that organized A Metaphysical Coming Together) and the Urantia Book study group offered a large experiential playground, but nothing like a Galactic Federation. I was incredibly curious at that point and wondered how my questions would be answered, if at all.

In June of 1988, I was sitting with a friend in our back yard in front of our guest house, which was being rented by him at the time. We'd gone to a vegetarian potluck and full-moon meditation earlier and, upon our return, fell into a deep conversation. We'd reached a point of stillness for a moment and I closed my eyes, looked inward and silently asked my proverbial question, "Who am I?"

Here comes the part that just made me stop and pay attention. Immediately I heard his chair move, opened my eyes to look at him and, locking on his eyes, I hear him say,

"You are Zendor." No precursive comment, just the answer.

Now if that didn't sound strange, I also had my second sight kick on. You know that imagery you get sometimes on your inner screen, and it often doesn't matter whether your eyes are open or closed? I saw a starscape laid out before me with a large stone-arched doorway and thick ornately carved wooden door in the forefront.

The door slowly opened, accompanied by the words (from somewhere else), "Door to what is." Then the whole scene vanished just as quick as it had appeared. I sat for a moment wondering what the heck to do. I shared it with my friend and left it alone. I used it for a book title several decades later to include what I thought was wisdom at the time. It would play an important role later.

Then my mind took over and started wanting to control the narrative. I really had some issues with that as I'd complained about people changing their names to some spiritual notion back in the 70s. I was not about to change my name. Still, I had to recognize that the answer to my question was delivered. I kept quiet for a while, unwilling to deal with its reality, let alone its implications.

A few months later while delivering the Arizona Light to bookstores around Arizona. I met a half-dozen or so 'psychics' at various bookstores and, after starting up conversations and developing rapport, I'd say something out of the blue, like, "Hey, do you mind if I ask a question?" They love questions.

What do you get from the name, Zendor?" The answers were even more confusing. Every one of them started with, "Oh, he's the commander of a mothership." It was freaking spooky. A few months later I started a discussion group on UFOs, with a couple of meeting locations in the east Phoenix valley.

One of the new members asked if we could meet for lunch,

that she had some important information she wanted to discuss. I was curious, though not naïve. There was no attraction, either. She did seem to have a 'different' energy than others in the group. There was a deeper awareness in her eyes that I recognized.

During our lunch, she began by recapping what she had perceived of me during our UFO discussion group. It was a bit more than just reading me as an NLP practitioner might. I asked her about the name and her answer sent me reeling. "He's a commander of a mothership and I know which one, the New Jerusalem." That one is still on the shelf.

Phase Four – Doris Gets Her Oats

I left the aerospace company in February of 1989 through disturbing circumstances and amidst severe depression for which I was self-medicating with pot at home. I was being granted asylum and didn't know it yet. The next year I collapsed into nothing, felt like I'd been burned to a cinder, and through a lot of soul searching got re-centered; rising from the ashes. I felt like I was a living example of the Phoenix.

Nineteen eighty nine was a year of deeper exploration and intense experience than I'd had since college. I was clean and clear, able to enter other states of consciousness I'd visited over a decade prior and retain more information and understanding. After staying in our guest house while another family moved into the house, I was ready to find another place to live. I happened quickly.

Almost as soon as I made the decision, I got a phone call from a member of one of the discussion groups I was leading, wanting to know if I knew anyone looking for a room in a metaphysical household. I moved that week and found a group that was working under the auspices of the Association for the Unfoldment of Man. Ray Stanford's work was their focus, along with the techniques developed by William Swygard, Multilevel Awareness and Multiplane Awareness.

I experienced my first and only personal 'channeling' ability where my perception of Jesus asked to speak through me after a group activity on a spiritual retreat at Woods Canyon Lake in northern Arizona. We were requested to 'act as if' Jesus was present and just *talk* to him.

It wasn't as easy as it may sound. Nearly everyone asked questions of how to rather than just speak up. After a while I piped up with, "Hey like this, *Hey Dude, glad to have you hear.*

Let's party!" The facilitator didn't seem to appreciate that and moved us on to the next activity quickly, a guided meditation, asking us to close our eyes and take a couple of deep cleansing breaths. That sounded intriguing as I'd gone on a few guided meditations a year prior.

As I closed my eyes and took a deep breath, this light appeared above me, so bright I had to look up. When I did, guess who? A beam of light from both his third eye and his heart combined about a foot and a half above me and bathed me with this effervescent iridescent feeling. I heard, "I would like to speak through you." I started crying instantly, with all kinds of feelings of unworthiness.

After I argued with myself for a few moments, reflecting on many voices from self and others while weeping uncontrollably, I decided I wasn't going to pass this opportunity up. I took a deep breath and let go, opened my mouth and out came these words, "Know that I am with you always. This one's fear is great." In my mind I knew the latter reference was toward me.

I got curious about why and he answered, like that thought separated my participant mind from the observer self and it became the participant. It was really weird, yet didn't impact the flow. I continued speaking to the group evidently from reports of the group afterward, "Your fears are the same as mine were." At that point I continued speaking, but was also engaged in a conversation and had no idea what I was saying to the group. I was completely absorbed in the conversation with Jesus as he said few words and mostly shared video, I suppose 'visions' would be another word.

A few minutes later, perhaps ten as I'm not sure how long exactly, I was brought back to the scene as I heard a branch snap as someone was walking up the trail nearby. I opened my eyes to see the entire group with their mouths gaping. I didn't say a word for hours afterward, only being able to hum instead and so enjoyed how the hum felt, too.

There was more and, not going further into detail, suffice it to say this Contact Modality left me humbled beyond belief. I was 31 at the time. Without any knowledge of what I'd said to the group, I was perplexed to say the least. I did understand how Steven and April didn't retain anything they had said, though. Interesting insight from the experience.

That summer I also went through a series of facilitated excursions into consciousness. The Multi-Level Awareness version was designed to open one to the awareness and initiation of the chakras, view Akashic records or work with one's spiritual guide. The Multi-Plane Awareness was designed to take the experiencer through 9 planes of consciousness with exercises on each to integrate the body on each plane. Information about both is available online.

No doubt there were a lot of moving parts to the system, yet there was definitely a system. I could *feel* it. I delved into more research, attending lectures and reading books as I attempted to discover the common threads I knew just had to be present. I found them later through a TV show I hosted.

I can't explain the knowing or why, only that it was a constant feeling of connectedness that led to a basic understanding of oneness, but it was still missing the science side of things and in the late 80s and early 90s, there wasn't much science-related material that I could find. The internet wasn't yet.

Dr. Jeffrey Mishlove was just beginning to expose those who were on the leading edge of science and the paranormal with a PBS television show called *Thinking Allowed, Conversations on the Leading Edge of Knowledge and Discovery*. In the back of my mind I hoped he'd interview me someday. I didn't even think I'd be hosting a show the following year.

I was captivated by Bill Moyers and Joseph Campbell through the conversations on the *Hero's Adventure* and *The Power of Myth*. Armed with the memory of the STE (Spiritually Transformative Experience) from college I found the archetypal representations a kind of play ground to explore the relationships that seemed to correlate, or at least offer another perspective of the event.

I got *A World of Ideas* books Bill authored and poured through the interviews. It created opportunities to explore the many of the greatest minds on the planet, with Bill asking the kinds of questions that I would want to ask and more. Being able to compare and contrast my own awareness and thinking was pure joy and challenged me to expand my world view even more as well as question my own perceptions.

I wanted to explore what others had to say as well, only from a more practical sense. Through yet another series of synchronistic events, fueled by intention and a wish years prior, I was presented with an opportunity to host a television show on Public Access. I had made a presentation to the Christown Lions Club in support of their help in creating the Phoenix part of The Earth Concert 1989. I hoped we could use their event insurance policy and access to their carnival.

The president was impressed and wanted me to meet a guy who was currently producing a show for them. We hit it off immediately, so much so that we spent a lot of time rock climbing throughout Arizona and Colorado over the next decade. I got trained for the studio work and participated as one of the crew, specifically on the Amiga graphics generator in the control room. It was a lot of fun and the actual work was minimal, so I could watch the control room activity, too.

Studio time was twice a month for three hours. Less than two months went by and the director got frustrated with the on-air talent, who was also wanting to move on. 'Harry the Origami Man' was tired of paper folding and had run out of stories, I guess. The director turned to me (the Amiga was right next to the switcher where he sat to direct and switch cameras) and asked if I'd be willing to do a show.

I was thrilled though I wasn't thrilled about his idea of doing a 'New Age' show because I had a lot of contacts within the metaphysical and new age community, having co-produced a large event called 'A Metaphysical Coming Together' at an iconic resort, known as the Biltmore. I knew anything close to 'new age' in the title wouldn't bode well, based on my interview on KTAR, although *One World* did fit.

The words just popped out of my mouth in discussion potential titles and they gave us pause. It had the least stigma of any other title we could come up with and was in total alignment with my 'mission.' I was bereft of format, though. I figured it would come, too, in its own timing. It did, during a meditation a few days later, unexpectedly. I suppose I should mention John Cleese speaks of this pattern for creatives.

The format came into my mind in sublime delivery, cascading through the questions complete with visuals. I started with questions toward guests about how they chose their professional path, what prompted them from both inner and outer perspectives. We explored the bridging of those perspectives and sharing their awareness of it.

We spent the bulk of the half-hour time investigating the fears they [guests] experienced and, more importantly, how they overcame them. This brought a sharing of their insights and understanding of their inner awareness and connectedness to the outer world in most cases. I interviewed actors, authors, business owners, doctors, lawyers, musicians, psychologists (animal, plant and human), story-tellers, vice-provosts and even politicians.

I also included contactees, healers, psychics and sensitives. Invariably, the process they went through was virtually the same although each individual experience of events were different. The same kinds of experiences were shared in a variety of articulations. The golden thread was the process of stepping through the fears. For nearly every guest, they'd come to know the fears as invitations to overcome. Some had a sense of the ineffable, the connectedness of all things that often appeared as synchronicities they recognized were like sign posts on the path to their success.

The show was my version of emulating both Mishlove and Moyers, initially hoping to someday be worthy of an interview with one of them. One World was an education like no other; a veritable Ph.D. on steroids in overcoming fear with perseverance and processes through faith, love and trust.

I still had questions about the orange cigar-shaped cloud I'd encountered. I didn't know where to find a person or publication that had any answers, let alone how to ask the right questions. I wasn't looking in the right places.

If there were things going on inside the cloud that caused me to have a sense of familiarity and desire to return, what were they? I know that early sense of *not* fearing the unknown, mostly because cool stuff was happening even if I didn't have a memory – I had a **sense**. It certainly put my future exploration in a different category than most with the ability to 'let go' learned from many angles.

Evidently the intention was fruitful. I wonder if it is all about 'attention, intention and interactions' with the world(s) that bring us knowledge, understanding and wisdom. I believe those will bring us harmony among people and planet... peace.

A Different Side of the Story

Late September in 1991, I woke up out of a sound sleep one night to find myself on what seemed like a bare steel gurney with only a T-shirt on. I sleep nude, so it was a bit strange right away (and I took notice). It all happened quickly, sitting up laughing as I noticed I had what looked like acupuncture needles arranged around my sphincter, with wires attached to a device sitting on a small table next to the wall directly in front of me. I was both embarrassed and intensely curious.

I do remember immediately saying, "Guys, what the heck do you think you are doing?" Pushed up against the wall next to the device were three tall grey (almond-eyed) beings with lab coats of sorts. I didn't know why, except that my laughter was probably completely unexpected. The needles were lining my sphincter and my laughter was more from embarrassment as I asked them what the heck they were doing while pulling them out one by one.

Later, after understanding more subtle energy and the difference in the size of our aura when afraid as opposed to being fully present and fearless, even laughing, I figured they were affected and responded from the 'push' of the 'energy' from the laughter, perhaps. At any rate, when I got to the last one I felt a slight sting, like that from accidently pulling a hair. It was short and sharp.

I woke up, yet again, back in bed. I immediately realized I just gave up a golden opportunity and just wanted to go back. The intention was really intense and yet with no understanding of 'how' to do it. I closed my eyes and 'tried' to go back with everything I could muster.

Instead of returning, a figure with a dark sparkling robe comes out of the darkness and says to me, "Next time, just relax. We're just trying to tune you up so we can communicate more easily." Wow, that was intense and at the same instant I understood the science behind the device, other dimensional as it was.

The perineum nerve extends from the tip of the spine and ends in the genitals and sphincter. If an intelligence of non-human origin understood anatomy and vibrational technology so well they were able to 'adjust' the frequency of a human being in such subtle ways as to open a greater experience of reality, well, it's the perfect plug in. The ecstasy would be intense initially, possibly producing an organism which might lead one to thing that sexual experiments were being done, although the intention was indeed much different.

The next few months there were several surreal events that had I not been a participant, I'd find them hard to believe. Even now as I write this I feel like I was just an observer, though I felt like a willing participant in an experiment so to speak. It was so spooky cool.

It started when I was asked to assist one of my guests from One World, part of a duo that claimed to remove etheric implants, Mary Margareht Rose and Royal David Brown. I had them examine me as a precursor to their interview, just to experience their process and figure out if they were legit or not. I was really skeptical, I must admit, though I had no preconceptions.

I didn't have any implants, but as I watched from within (eyes closed) I could see what looked like laser lights coming from their eyes and going through the backs of their hands as they moved them across my body just a few inches above it.

After I saw this, and with no words spoken, I had the undeniable urge to unfurl my hands (they were closed) and 'send' a bolt of energy up through their bodies and upward toward the sky. I didn't think anything beyond that.

As I did, both of them simultaneously stood straight up for a moment and then bent back over to continue their 'work.' At that point I spoke up, asked them what they had just experienced and waited for the answer. They both felt like they had been 'taken up' to a ship for an instant and then returned. I have no way of explaining what happened or even if it did, only to report my subjective experience of the event and their response.

The call was another matter entirely, as the woman was a bit frantic and claimed there was a tall grey hunched over as if injured, just outside the door to their office. It used to be a maintenance apartment for the office building it was behind, sharing a common wall yet a perfect space for them with several rooms.

She said he was obviously hurt, but she had difficulty communicating with him (she claimed to have regular interactions with ETs when we first met) and thought that maybe it was a gender thing. The overpowering thought was to call me and I'd know what to do. I had that skepticism come up yet again.

It was early afternoon and after considering that it could all just be fantasy, I decided I could prove she was wrong. I would enjoy the opportunity of it being true, but not without healthy skepticism. It was only about a 15 minute journey and on the way there I inquired if it really was real or not. Instantly it felt like a javelin struck my hip joint; the pain so great I had to pull over for a moment. I didn't know what to think or do, but I figured it would be interesting.

When I arrived I was invited into the 'healing room' which had a massage table in the center and various metaphysical accoutrements. I'd been in there for my examination. I had no line of sight into the room from the door, but as I entered the room I blinked and was presented a startling image.

I opened my eyes again to see nothing on the table, yet with my eyes closed I'd just seen a reptilian, a Draconian as I've come to know, in a body-tight uniform with some insignias on its breast. His legs and feet extended several feet beyond the end of the table. I didn't have my eyes closed long enough to garner further detail. I didn't want to look again, though.

It freaked me out, yet I wasn't freaked out. In that moment it felt like another part of me stepped in to perform, so I just let go of any fear or resistance and moved from within. I knew I saw him, that wasn't a question. I was more concerned about why I was there and obviously connected to him somehow.

I directed Mary to stand at the end of the table as I moved a stool over to sit above his head, then placed my hand as if I were putting them on top of his head and imagined sending pure energy through it. I had her stand at his foot to receive the energy and direct it into the earth; an intentional act more than physical.

As soon as convenient I asked her if she'd examined him and found his injury. She did. It was his left hip, exactly where I'd felt the pain. She has established a telepathic link and I wanted no part of it. I felt that I had to remain free of it for some reason, like it was a grounding affect for me not to get involved.

I asked her questions and got answers, pretty weird ones. He was a commander of three ships, who we could call Hurley, and had been tasked with watching me and over some years. Through this observation he and his crew had gone from being part of the group that was responsible for the implants to creating a rebellion and destroying the 'facility' where the implantations took place.

His injury was a result and it was time for him to reveal

himself, knowing I was the only one that could grant them permission to join the Galactic Federation. That was a bit much for me to take on one hand, and on the other I found myself making arrangements telepathically with the 'crew.' That's incredulous, huh? Part of me felt bat shit crazy and yet the majority felt completely lucid.

Other events included a pair of reptilians and insectoid-looking beings appearing in my bedroom around a small table. I can't imagine what their true size was, but they fit into an area as some kind of holographic extension or projection. I saw them with my eyes closed first, after hearing/feeling an extremely high- pitched sound inaudible to my physical ears, or so I thought. I felt it, regardless, and it was rather intense.

I didn't know whether to open them or not, but eventually I did and saw the same thing. I'd had experiences as a teen where I saw the same thing with eyes opened and closed, so it didn't totally freak me out. That part of me that was present with Hurley was watching, though. The four gave reports of progress that I didn't fully understand, yet did feel compelled to verify that no malevolence toward or manipulation of humans would be involved.

The conversation may have lasted a few minutes, if that, and then the same ringing and they were gone. I was even more puzzled than before, even though there was some history of this apparent theme that involves my interactions with some Galactic Federation and its local representation in the Ashtar Command. As wacky as it sounds, there was something going on deep inside me that I simply could not deny, no matter how hard I tried.

I'm still not sold on the idea, but this all started with George Adamski and then George Van Tassel back in the 50s and I suppose it's always possible some kind of evolving engagement has been happening. I believe both articulated their experience as well as anyone could, based on the approach of those whom they encountered over a half-century ago.

There's much more, and perhaps I'll include it in a future book. I've already written several, with *Stubbing My T.O.E. on Purpose* getting an interview with Dr. Mishlove in August of 2018, a 26-year dream come true. He reached out to me to share an interview with Jason Jorjani on the Greys for UfologyPress. I reviewed it and sent the link, along with a copy of my book.

I got an email back inviting me to come and talk about it. I was absolutely stunned and thrilled simultaneously, sharing the message with my wife, Luba. I introduced Jeffrey to FREE as well and mentioned Beyond UFOs during our interview. He wasn't aware of the organization yet. Later he was able to interview Rey and I'm sure the work of FREE greatly benefitted as a result.

I'm still not sure about the depth of this non-human intelligence as what I've shared indicates a plethora of possibilities, and what other kinds of beings even beyond those of the Galactic Federation may be looking or lurking. The Urantia Book indicates that is just the tip of the proverbial iceberg, so who knows what we'll actually discover in the next half-century.

One of the key points of our interviews was a statement I made about our bodies, that they are *essentially an instrument that we simply haven't learned to play well*, yet. I will say that the consistent and resident theme across the dimensions I've experienced is harmony and the desire for us is that we'll experience harmony among people and planet as soon as possible. They're all, in their many splendored forms and contrary to conspiratorialists, here to help humanity rise to the occasion.

Relationships via the Contact Modalities?

Pieces of Our Holistic System

There are many models and theories of consciousness that attempt to explain the relationships of human dynamics, our ability to perceive and ruminate on how we interact with our reality. In a broad brushstroke of inclusion, all of them are unsatisfactory in providing an explanation for the transfer of information from sources beyond the physical, yet are undeniably a part of our human experience and often embedded in it.

Subtle energy or nearly imperceptible vibrations around us always, perceptively available in mentally 'still' place, offers greater intuitive capacity. Non-human intelligence often seems to be occupying this same space, so are we intuiting or perceiving from another source? I don't think we know, yet, though the indications of a far greater capacity for communication are right in front of us.

It would seem that to know ourselves first, the nature of how we function on as many levels as possible, would give us the ability to understand the relationship between our consciousness, cosmology and the interaction with others. There are a number of built in systems we have access to and understanding of from the ancient arcane schools of mysticism.

Too often we look forward without the proper foundation from which to achieve the best view. If we are uneducated about the nature and understanding of 'self,' then we are often mislead by those who seek to redirect our thinking into fear-based propaganda that serves no one except those who seek to control. It doesn't seem that non-human intelligence, in any way, seeks to control or manipulate humans.

In order to have a little 'common' sense I'd like to explore the basic systems of intuition and sensitivity that have been around for a long time. Our society, especially in the West, has downplayed and often ridiculed the quest of the spiritual seeker to align with the 'flow' of creation. Instead, we've subjugated the 'will' to serve needs other than the good of all.

The latter appears to be a central theme in the development of this understanding. It is both personal and global. The Law of One refers to the polarity of 'service to self' and 'service to others' as the challenge we face individually and collectively. Extrapolating the meaning indicates the service to others as being the ultimate service to self, the real personal needs being met in the process. One of the most famous quotes from Zig Ziglar illustrates the concept that "You can have everything in life you want, if you will just help other people get what they want."

Physiological Instrumentation

So if we are indeed an instrument, then how do we learn how to 'play' it? What do we know about it? Early on, after visiting the psychiatrist, I became interested in understanding energy, especially that of the physical body and its natural systems, and the implications of what were known as Near-Death experiences. They were the only 'similar' experiences I could find as I researched an explanation for my own experience.

Dividing my time between school study and personal study became a challenge and my studies for school suffered greatly. I was on a mission to explore me, not the biology and chemistry for my Pre-med program. My second year I changed my major to Religion with a Philosophy minor and fell into an abyss of self- exploration, part of which included the drums I purchased, and experimentation beyond reasonable notions.

There are certainly many more aspects than I can discuss here. I've probably forgotten more than I've retained, yet somewhere in my brain files, the information is stored and, when necessary, if often emerges in the moment. I think the brain is an amazing device, capable of storing and offering up information with precision.

We don't always know just how to access the files. Just to help us get on the same page I'll present some information in the following paragraphs on 'psychic' senses (I call them the 'clair-alls'), the chakra system, meridian system and the Multi-Plane Awareness system. This certainly does not constitute a complete list, only a few systems that I'll use for reference.

The **psychic senses** all begin with the suffix of 'clair' which means 'clear.' The **chakra system** originates from East Indian spiritual practice and the **meridian system** from Chinese traditional medicine. For purposes of identification and sharing information to make sense common in some way, I'll share their names and definitions below. *(Sources Identified)* Two additional 'models' of experiential opportunity are the Multi-Level and Multi-Plane Awareness techniques developed by William Swygard in the 1950s.

Clair- Senses

(from: http://www.okinhealth.com/articles/10-clairsenses-intuition-emily-matweow)
Clair *is a word describing types of clear sensitivity corresponding to our physical and intuitive senses. Clair begins words that name our intuitive abilities. The following are definitions of the various Clair Senses:*

> *CLAIRCOGNIZANCE – INTUITIVE KNOWLEDGE*
>
> *Clear knowledge is when a person has psychic knowledge without any physical explanation or reason. Claircognizance includes precognition and retro cognition – knowledge of the future and knowledge of the past. There are no restrictions as to*

what may be known with clear knowledge.
CLAIRVOYANCE – INTUITIVE VISION

The ability to see objects, actions, or events distant from the present without the use of eyes is called clairvoyance. It transcends time and space and may be a result of what is seen through the third eye (sometimes referred to as our mind's eye) or the fourth eye.
CLAIRAUDIENCE – INTUITIVE AUDIO or HEARING

This is the ability to perceive sounds or words and extrasensory noise from sources broadcasting from the spiritual or ethereal realm. These tones exist beyond the reach of rational human experience and beyond the limitations of ordinary space and time. For example:

- sounds made by a person's body;
- sounds made by living things;
- sounds made by nature;
- sounds made by man-made things;
- sounds made by interactions of the above; and
- ethereal sounds like voices of the dead, specters, or mystic music.
CLAIRSENTIENCE – INTUITIVE KNOWING BY FEELING

This refers to a person's ability to acquire knowledge by feeling. A person who feels the vibration of other people, animals, and places is clairsentient. There are many degrees of clairsentience, ranging from the perception of thoughts and emotions in others to their illnesses and injuries. This ability differs from clairvoyance because the knowledge comes only from feeling in the body.
Clairsentience includes an individual feeling the physical and emotional pain of land where atrocities have happened; jealousy, insecurity, fear, or displeasure in others; or others' physical pain.
CLAIRSALIENCE - INTUITIVE SMELL

Clairsalience, also known as clairscent or clairscentency, involves smelling a fragrance or odor of a substance, person, place, or animal not in one's surroundings. These odors are perceived without the use of the physical nose and beyond the limitations of ordinary time and space.

CLAIRTANGENCY INTUITIVE KNOWING by TOUCHING

Clairtangency is psychometry. The clairtangent handles an object or touches something and in doing so knows information about the object or its owner or its history that was not known beforehand. Clairtangency can also apply to touching a living being.

CLAIRTACTION – INTUITIVE TOUCH

I introduce this label for a "new" unnamed Clair. I chose taction because it is an archaic word defined as "the act of touching or making contact."

Clairtaction is the ability to sense being touched by a spiritual being or entity and the knowing of information about that spirit. Further, it includes a telekinetic-like ability to extend a touch to both physical and etheric entities in such a way that both the recipient and the psychic have awareness of the feeling. This means that the clairtactient may touch a person who is present or remote without physically touching him or her and the person being touched will, if aware enough, be able to identify where he or she is being touched and the nature of the touch. It also means that the clairtactient can touch something ethereal with both purpose and awareness.

CLAIRGUSTANCE – INTUITIVE TASTE

Clear tasting involves being able to taste something without actually putting it into your mouth. It's the perfect diet plan.

CLAIREMPATHY – INTUITIVE FEELING of EMOTION

Clairempathy is the ability to know people and their energies. A person who has clairempathy psychically experiences the thoughts or attitudes of a person, place, or animal and then feels the associated mental, emotional, physical, and/or spiritual results. The result of this clear empathetic experience may be a very physical response for the psychic in proportion to the

psychic's empathic sensitivity.
CLAIRELOQUENCE – INTUITIVE COMMUNICATING

Claireloquence is a second new Clair. It is the requirement to use precisely the right word or combination of words in order to accomplish a specific objective. Claireloquence connotes that it is the exact meaning of a word that is the result when it may in fact be the performative nature of the word or the exact phonology, or sound system, of the word to encode meaning that is the desired result. I believe that it can be either and both.
CLAIRESSENCE – INTUITIVE EMBODIMENT

Clairessence is a third new Clair I have come to understand and validate. It is the fundamental "Clair of all Clairs." It is both the easiest and the most difficult to explain, for it is the Clair of Ascension; by the time we fully understand this Clair it will no longer be significant because we will be "it".

The origin of clairessence is the point of origin of our nature. It is the mathematical and vibrational perfection with which all life was created. It is the ideal we seek to attain in the successful embodiment and integration of all of our senses within our rational and non-rational selves. It is the state we seek to return to – a state without distortion that is more refined, by virtue of experience, than when it was created.

Chakra System
(from: http://www.chakras.info/7-chakras/)
The study of 7 chakras originates in Eastern spiritual traditions that consider the seven primary chakras the basis of our human existence. Similarly, today's Western approaches place an emphasis on the seven chakras as representations of different aspects of our life and describe their function in various terms encompassing the psychological, physical, energetic and spiritual.

7 Major Chakras Overview
The basic human chakra system, as it is commonly accepted, consists of seven chakras stretching from the base of the spine to the crown of the head. Their names, locations and corresponding

chakra colors are:

1. ***Root chakra** — base of the spine — red*
2. ***Sacral chakra** — just below the navel — orange*
3. ***Solar Plexus chakra** — stomach area — yellow*
4. ***Heart chakra** — center of the chest — green*
5. ***Throat chakra** — base of the throat — blue*
6. ***Third Eye chakra** — forehead, just above area between the eyes —
indigo*
7. ***Crown chakra** — top of the head — violet*

Sometimes, a Sanskrit name is used instead of plain English because the study of the chakra system as we know it in our modern Western culture originates mostly from yogic traditions from India. You'll sometimes find them referenced as:

1. *"Muladhara"*
2. *"Svadhishthana"*
3. *"Manipura"*
4. *"Anahata"*
5. *"Vishuddha"*
6. *"Ajna"*
7. *"Sahasrara"*

Other references of the mid-20th century include as many as 13 chakras, with the other 6 residing above and below the physical body, including the 'cosmic' and 'earthly' energy bodies.

Chinese Traditional Medicine Meridian System –
(From: https://kootenaycolumbiacollege.com/the-chinese-medicine-meridian-system/)

So, what is a meridian anyway? This is one of the first questions students of Chinese medicine want to understand. Simply put, a meridian is an 'energy highway' in the human body. Qi (chee) energy flows through this meridian or energy highway, accessing all party of the body. Meridians can be mapped throughout the body; they flow within the body and not on the surface, meridians exist in corresponding pairs and each meridian has many acupuncture points along its path.

The term 'meridian' describes the overall energy distribution system of Chinese Medicine and helps us to understand how basic substances of the body (Qi, blood and body fluids) permeate the whole body. The individual meridiansthemselves are often described as 'channels' or even 'vessels' which reflects the notion of carrying, holding, or transporting qi, blood and body fluids around the body.

It is tempting to think of the meridians of the human body the same way as we think of the circulatory system, as the meridians are responsible for the distribution of the basic substances throughout the body just like the circulatory system, but here is where the similarities end. Conventional anatomy and physiology would not be able to identify these pathways in a physical sense in the way that blood vessels can be identified.

It is more useful to consider the meridian system as an energetic distribution network that in itself tends towards energetic manifestation. Meridians can be best understood as a process rather than a structure.

Practitioners of Chinese Medicine must be as knowledgeable about these meridian channels as the Western Doctor is about anatomy and physiology of the physical body. Without this thorough understanding, successful acupuncture treatments would be difficult. A practitioner of Chinese Medicine must know how and where to access the qi energy of the body to facilitate the healing process.

There are twelve main meridians, or invisible channels, throughout the body with Qi or energy flows. Each limb is traversed by six channels, three Yin channels on the inside, and three Yang channels on the outside. Each of the twelve regular channels corresponds to the five Yin organs, the six Yang organs as well as the Pericardium and San Jiao. These are organs that have no anatomical counterpart in Western medicine but also relate to processes in the body. It is also important to remember that organs should not be thought of as being identical with the physical, anatomical organs of the body.

Each meridian is a Yin Yang pair, meaning each Yin organ is paired with its corresponding Yang Organ: the Yin Lung organ, for example, corresponds with the Yang large intestine.

Qi flows in a precise manner through the twelve regular meridians or channels. First, Qi flows from the chest area along the three arm Yin channels (Lung, Pericardium, and Heart) to the hands. There they connect with the three paired arm Yang channels (Large Intestine, San Jiao and Small Intestine) and flow upward to the head. In the head they connect with their three corresponding leg Yang Channels (Stomach, Gall Bladder and Bladder) and flow down the body to the feet. In the feet they connect with their corresponding leg Yin channels (Spleen, Liver, Kidney) and flow up again to the chest to complete the cycle of Qi.

- *Arm Tai Yin channel corresponds to the Lung*
- *Leg Tai Yin channel corresponds to the Spleen*
- *Arm Shao Yin channel corresponds to the Heart*
- *Leg Shao Yin corresponds to the Kidney*
- *Arm Jue Yin corresponds to the Pericardium*
- *Leg Jue Yin corresponds to the Liver*
- *Arm Yang Ming corresponds to the Large Intestine*
- *Leg Yang Ming corresponds to the Stomach*
- *Arm Tai Yang corresponds to the Small Intestine*
- *Leg Tai Yang corresponds to the Bladder*
- *Arm Shao Yang corresponds to the San Jiao*
- *Leg Shao Yang Channel corresponds to the Gall Bladder*

The arm and leg channels of the same name are considered to 'communicate' with each other in Chinese medicine. Thus, problems in a given channel or organ can be treated by using various points on the communication 'partner'. As an example: a problem with the lungs can be treated by using points on the Spleen channel as they are both Tai Yin channels.

In addition to the twelve regular meridians there are 'Extraordinary Meridians' that are not directly linked to the major organ system but have various specific functions:

1) reservoirs of Qi and blood for the twelve regular channels, filling and emptying as required
2) circulate jing or 'essence' around the body because they have a strong connection with the Kidneys

3) circulate the defensive Wei Qi over the trunk of the body and, as such, play an important role in maintaining of good health

4) provide further connections between the twelve regular channels

The meridian system of the human body is a delicate, yet intricate web of interconnecting energy lines. If a person masters an understanding of this meridian system they will know the secrets of the flow of Qi energy in the body.

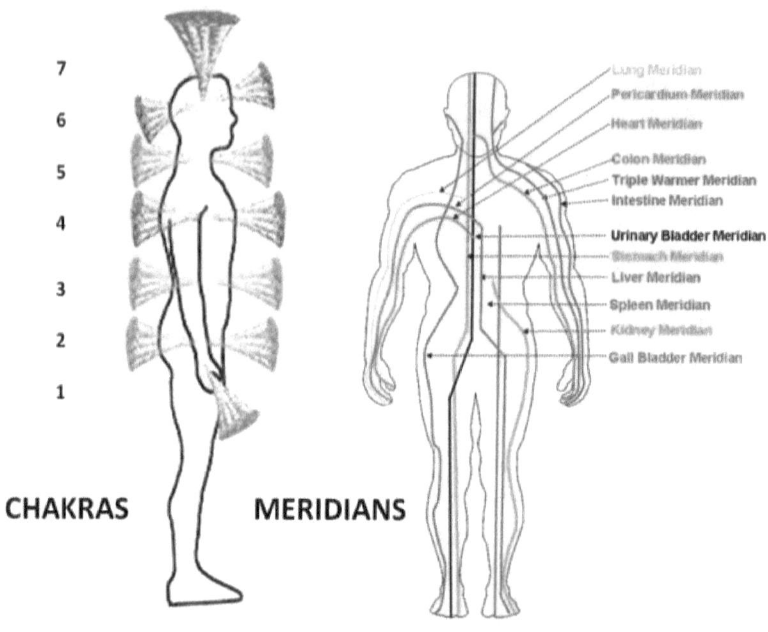

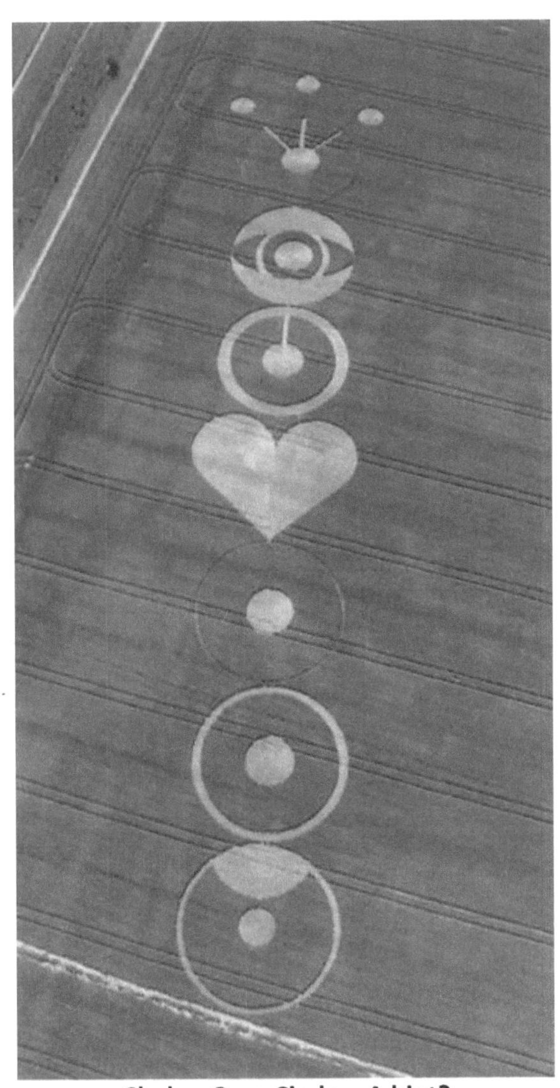

Chakra Crop Circle – A hint?

Marma Points of Ayurveda

Marma points are energy points in the body used for healing in Ayurveda. They can be compared to acupuncture points and meridians in Chinese Medicine.

Literally translated, marma means 'a point that can kill', and indeed some of the marma points have been identified and used in martial arts, however in marma point massage, these points are only used for healing purposes! They're also identified as neurolymphatic points, stimulating the removal of lymph and enhancing the efficiency of the body's organs.

A marma point is a juncture on the body where two or more types of tissue meet, such as muscles, veins, ligaments, bones or joints. Although Marma points are much more than a casual connection of tissue and fluids, they are points of the vital life force.

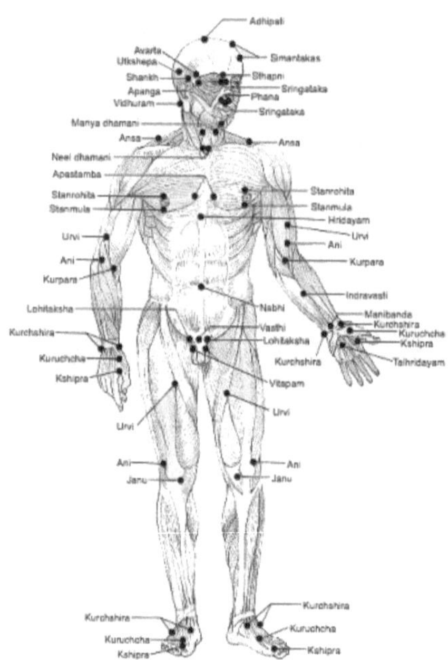

Multi-Level Awareness

A much more recent introduction of esoteric opportunity

came in the mid-20th century with William Swygard's work. He had a unique understanding of assisting others to journey inward. Although rarely credited or mentioned, nearly all the journeying techniques today have some element of his work.

> The Multi-Level Awareness technique was quite unique in its usefulness across a number of metaphysical applications. Through the expertly facilitated process I was able to explore four basic areas of the self; engaging the Akashic records, chakras, past lives and spirit guides weren't new, yet the process was structured and repeatable.
>
> In order to engage any of the threads mentioned, the initial process included a series of directed inner scenes that resolved into a freedom of movement into the specific area of exploration. In most cases only one of the options was explored in order to maintain focus and have an optimal experience. With a trained facilitator, not only was the journey productive, there were notes of the observational responses of the experiencer. Further investigation of the experiences, as non-linear and non-local events, was made possible by the captured reporting. This was all done long hand as technology for recording was limited early on.

Multi-Plane Awareness

This technique is the most unique I've found and perhaps the most profound in experiential observation. It combined the elegance of astral travel, vibrational structure of the electromagnetic spectrum and an understanding of quantum mechanics and multiple planes of consciousness we inhabit.

For me, it put an experience to the notion of a multidimensional reality that I'd experienced throughout my lifetime to date. I was introduced to the process in 1989, so I've had many years to distill the experiences and garner at least some understanding as well as pass it on to others.

Expressing the understanding is a challenge, so bear with me.

Suffice it to say we start with the 1st plane as the Void, so it has been called, with the 2nd plane being one of light surrounding the Void, nearly infinitesimally small and indistinct. The 3rd plane is our physicality, the 4th the first astral and on up the scale in higher vibrational ranges that also correspond with an expanding awareness of layers of consciousness that intertwine across the electromagnetic spectrum. It seemed as though each plane become more expansive and less personal, merging into a larger collective with a unifying theme and seeming understanding of universe administration that is ineffable.

Below are some additional considerations:

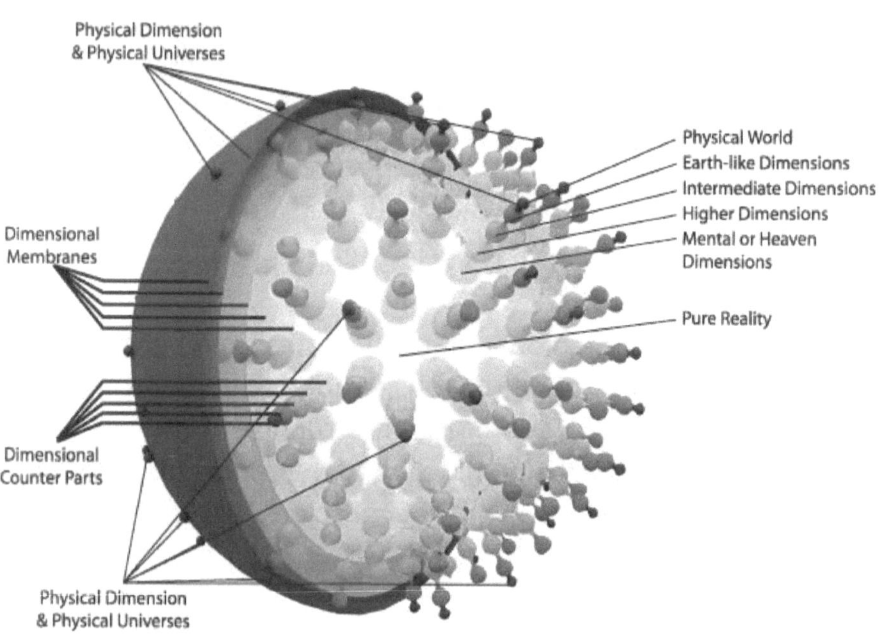

Another Possibility

Many are accustomed to the toroidal magnetic field as a consideration for the energy field the body emits, or any cosmic body for that matter. Based on the information provided so far, consider innumerable energetically sensitive tendrils going out and in with each heartbeat, primarily through the chakras and meridians, like a self-generated pulsing magnetic field, sharing and receiving information constantly.

Though the inclusion of the above is certainly perceivable as simultaneous actions, the depth of awareness necessary to experience the events in the body rarely occurs. Even the notion of it can have profound effects on one's experience in the world. The condensation of consciousness in the body is our first entrance into the world at birth.

So the body is how the mind interacts, yes, inextricably connected, and yet the body is the primary data collector and translates naturally in the brain when allowed to do so. Thinkers that we are, this process usually gets sidetracked by our distractions of attention.

However, we can override that through the use of free will and learning how to make appropriate choices. There's no manual on that, though there are many instructional books and videos as well as individuals who've at least mastered some of the process. Alas, ignoring the sensations from the body is an option always. Whether we perceive (become aware) or not, the *system* is always working.

The Human Toroidal Field

Humans, both as individuals and in groups, also are surrounded by a toroidal field. The axis of the field passes through the crown of the head and out the center of the groin. The strength of the field is directly related to the vitality of the person it surrounds. Each person's torus in unique but also open to interconnecting with the fields of others.

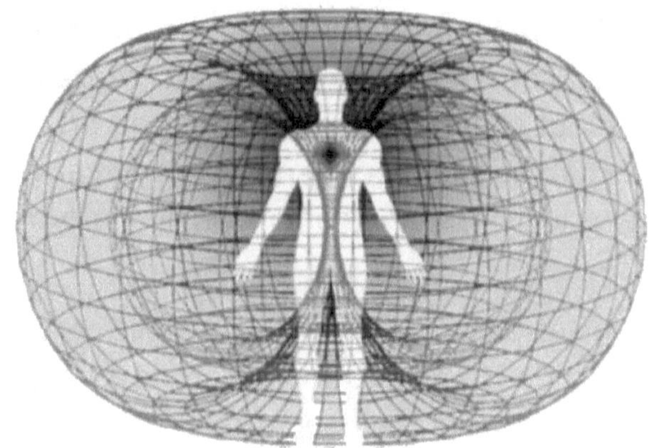

The combined strength of human torus fields connecting together is best demonstrated when attending events or parties. . Our energy fields can create a sub-conscious kind of telepathy.

The human body which consists mostly of water is very affected by sound and vibration. This, in turn, can amplify the toroidal field surrounding the body and, in groups, create a sort of human "wifi" as the tori grow and connect. This stimuli and combination effect give the group much stronger power both as individuals and especially overall.

The deep, spiritual feeling of interconnectedness is extremely addictive and powerful. All emotions and feelings are infinitely amplified as others combine their energy to the group feeling. This helps us to understand the incredible popularity of concerts and sporting events as well as the effect of mob mentality. Anger can quickly become rage, happiness can become euphoric joy.

666
999

Cosmological Theories

How much do we really know about our cosmology? There are many models and theories and I've included most of them here, the advantage of doing a little research and crediting the source. I'll have an interesting perspective to offer in answer to the final question. Many experiencers may not know how or where to look for additional information about the various cosmological theories, nor do I know if they'll be included in the other chapters, so I'm including some information here that was valuable to me. Perhaps you will find value as well.

From: **(https://www.physicsoftheuniverse.com/cosmological.html)**

"Cosmos" is just another word for universe, and "cosmology" is the study of the origin, evolution and fate of the universe. Some of the best minds in history - both philosophers and scientists - have applied themselves to an understanding of just what the universe is and where it came from, suggesting in the process a bewildering variety of theories and ideas, from the Cosmic Egg to the Big Bang and beyond. Here are some of the main ones, in approximate chronological order:

Brahmanda (Cosmic Egg) Universe - The Hindu Rigveda, written in India around the 15th - 12th Century B.C., describes a cyclical or oscillating universe in which a "cosmic egg", or Brahmanda, containing the whole universe (including the Sun, Moon, planets and all of space) expands out of a single concentrated point called a Bindu before subsequently collapsing again. The universe cycles infinitely between expansion and total collapse.

Anaxagorian Universe - The 5th Century B.C. Greek philosopher Anaxagoras believed that the original state of the cosmos was a primordial mixture of all its ingredients which existed in infinitesimally small fragments of themselves. This mixture was not entirely uniform, and some ingredients were present in higher concentrations than others, as well as varying from place to place. At some point in time, this mixture was set in motion by the action of "nous" (mind), and the whirling motion shifted and separated out the ingredients, ultimately producing the cosmos of separate material objects, all with different properties, that we see today.

Atomist Universe - Later in the 5th Century B.C., the Greek philosophers Leucippus and Democritus founded the school of Atomism, which held that the universe was composed of very small, indivisible and indestructible building blocks known as atoms (from the Greek "atomos", meaning "uncuttable"). All of reality and all the objects in the universe are composed of different arrangements of these eternal atoms and an infinite void, in which they form different combinations and shapes.

Aristotelian Universe - The Greek philosopher Aristotle, in the 4th Century B.C., established a geocentric universe in which the fixed, spherical Earth is at the center, surrounded by concentric celestial spheres of planets and stars. Although he believed the universe to be finite in size, he stressed that it exists unchanged and static throughout eternity. Aristotle definitively established the four classical elements of fire, air, earth and water, which were acted on by two forces, gravity (the tendency of earth and water to sink) and levity (the tendency of air and fire to rise). He later added a fifth element, aether, to describe the void that fills the universe above the terrestrial sphere.

Stoic Universe - The Stoic philosophers of ancient Greece (3rd Century B.C. and after) believed in a kind of island universe in which a finite cosmos is surrounded by an infinite void (not dissimilar in principle to a galaxy). They held that the cosmos is in a constant state of flux, and pulsates in size and periodically passes through upheavals and conflagrations. In the Stoic view, the universe is like a giant living body, with its leading part being the stars and the Sun, but in which all parts are interconnected, so that what happens in one place affects what happens elsewhere. They also held a cyclical view of history, in which the world was once pure fire and would become fire again (an idea borrowed from Heraclitus).

Heliocentric Universe - The 3rd Century B.C. Greek astronomer and mathematician Aristarchus of Samos was the first to present an explicit argument for a heliocentric model of the Solar System, placing the Sun, not the Earth, at the center of the known universe. He described the Earth as rotating daily on its axis and revolving annually about the Sun in a circular orbit, along with a sphere of fixed stars. His ideas were generally rejected in favor of the geocentric theories of Aristotle and Ptolemy until they were successfully revived nearly 1800 years later by Copernicus. However, there were exceptions: Seleucus of Seleucia, who lived about a century after Aristarchus, supported his theories and used the tides to explain heliocentricity and the influence of the Moon; the Indian astronomer and mathematician Aryabhata described elliptical orbits around the Sun at the end of the 5th Century A.D.; as did the Muslim astronomer Ja'far ibn Muhammad Abu Ma'shar al-Balkhi in the 9th Century.

Ptolemaic Universe - The 2nd Century A.D. Roman-Egyptian mathematician and astronomer Ptolemy (Claudius Ptolemaeus) described a geocentric model largely based on Aristotelian ideas, in which the planets and the rest of the universe orbit about a stationary Earth in circular epicycles. In terms of longevity, it was perhaps the most successful cosmological model of all time. Modifications to the basic Ptolemaic system were suggested by the Islamic Maragha School in the 13th, 14th and 15th Centuries including the first accurate lunar model by Ibn al-Shatir, and the rejection of a stationary Earth in favor of a rotating Earth by Ali Qushji.

Abrahamic Universe - Several medieval Christian, Muslim and Jewish scholars put forward the idea of a universe which was finite in time. In the 6th Century A.D., the Christian philosopher John Philoponus of Alexandria argued against the ancient Greek notion of an infinite past, and was perhaps the first commentator to argue that the universe is finite in time and therefore had a beginning. Early Muslim theologians such as Al-Kindi (9th Century) and Al-Ghazali (11th Century) offered logical arguments supporting a finite universe, as did the 10th Century Jewish philosopher Saadia Gaon.

Partially Heliocentric Universe - In the 15th and early 16th Century, Somayaji Nilakantha of the Kerala school of astronomy and mathematics in southern India developed a computational system for a partially heliocentric planetary model in which Mercury, Venus, Mars, Jupiter and Saturn orbited the Sun, which in turn orbited the Earth. This was very similar to the Tychonic system proposed by the Danish nobleman Tycho Brahe later in the 16th Century as a kind of hybrid of the Ptolemaic and Copernican models.

Copernican Universe - In 1543, the Polish astronomer and polymath Nicolaus Copernicus adapted the geocentric Maragha model of Ibn al-Shatir to meet the requirements of the ancient heliocentric universe of Aristarchus. His publication of a scientific theory of heliocentrism, demonstrating that the motions of celestial objects can be explained without putting the Earth at rest in the center of

the universe, stimulated further scientific investigations and became a landmark in the history of modern science, sometimes known as the Copernican Revolution. His Copernican Principle (that the Earth is not in a central, specially favored position) and its implication that celestial bodies obey physical laws identical to those on Earth, first established cosmology as a science rather than a branch of metaphysics.

In 1576, the English astronomer Thomas Digges popularized Copernicus' ideas and also extended them by positing the existence of a multitude of stars extending to infinity, rather than just Copernicus' narrow band of fixed stars. The Italian philosopher Giordano Bruno took the Copernican Principle a stage further in 1584 by suggesting that even the Solar System is not the center of the universe, but rather a relatively insignificant star system among an infinite multitude of others. In 1605, Johannes Kepler made further

refinements by finally abandoning the classical assumption
of circular orbits in favor of elliptical orbits which could
explain the strange apparent movements of the planets.
Galileo's controversial support of Copernicus' heliocentric
model in the early 17th Century was denounced by the
Inquisition but nevertheless helped to popularize the idea.

Cartesian Vortex Universe - In the mid-17th Century, the
French philosopher René Descartes outlined a model of the
universe with many of the characteristics of Newton's later
static, infinite universe. But, according to Descartes, the
vacuum of space was not empty at all, but was filled with
matter that swirled around in large and small vortices. His
model involved a system of huge swirling whirlpools of
ethereal or fine matter, producing what would later be called
gravitational effects.

Static (or Newtonian) Universe - In 1687, Sir Isaac Newton
published his "Principia", which described, among other things,
a static, steady state, infinite universe which even Einstein, in
the early 20th Century, took as a given (at least until events
proved otherwise). In Newton's universe, matter on the large
scale is uniformly distributed, and the universe is gravitationally
balanced but essentially unstable.

Hierarchical Universe and the Nebular Hypothesis - Although
still generally based on a Newtonian static universe, the matter
in a hierarchical universe is clustered on even larger scales of
hierarchy, and is endlessly being recycled. It was first proposed
in 1734 by the Swedish scientist and philosopher Emanuel
Swedenborg, and developed further (independently) by
Thomas Wright (1750), Immanuel Kant (1755) and Johann
Heinrich Lambert (1761), and a similar model was proposed in
1796 by the Frenchman Pierre-Simon Laplace.

Einsteinian Universe - The model of the universe assumed by
Albert Einstein in his groundbreaking theory of gravity in the
early 20th Century was not dissimilar to Newton's in that it
was a static, dynamically stable universe which was neither

expanding or contracting. However, he had to add in a " cosmological constant" to his general relativity equations to counteract the dynamical effects of gravity which would otherwise have caused the universe to collapse in on itself (although he later abandoned that part of his theory when Edwin Hubble definitively showed in 1929 that the universe was not in fact static).

Big Bang Model of the Universe - After Hubble's demonstration of the continuously expanding universe in 1929 (and especially after the discovery of cosmic microwave background radiation by Arno Penzias and Robert Wilson in 1965), some version of the Big Bang theory has generally been the mainstream scientific view. The theory describes the universe as originating in an infinitely tiny, infinitely dense point (or singularity) between 13 and 14 billion years ago, from where it has been expanding ever since. The essential statement of the theory is usually attributed to the Belgian Roman Catholic priest and physicist Georges Lemaître in 1927 (even before Hubble's corroborating evidence), although a similar theory had been proposed, although not pursued, 1922 by the Russian Alexander Friedmann in 1922. Friedmann actually developed two models of an expanding universe based on Einstein's general relativity equations, one with positive curvature or spherical space, and one with negative curvature or hyperbolic space.

Oscillating Universe - This was Einstein's favored model after he rejected his own original model in the 1930s. The oscillating universe followed from Alexander Friedmann's model of an expanding universe based on the general relativity equations for a universe with positive curvature (spherical space), which results in the universe expanding for a time and then contracting due to the pull of its gravity, in a perpetual cycle of Big Bang followed by Big Crunch. Time is thus endless and beginningless, and the beginning-of-time paradox is avoided.

Steady State Universe - This non-standard cosmology (i.e. opposed to the standard Big Bang model) has occurred in various versions since the Big Bang theory was generally adopted by the scientific community. A popular variant of the steady state universe was proposed in 1948 by the English astronomer Fred Hoyle and the Austrians Thomas Gold and Hermann Bondi. It predicted a universe that expanded but did not change its density, with matter being inserted into the universe as it expanded in order to maintain a constant density. Despite its drawbacks, this was quite a popular idea until the discovery of the cosmic microwave background radiation in 1965 which supported the Big Bang model.

Inflationary (or Inflating) Universe - In 1980, the American physicist Alan Guth proposed a model of the universe based on the Big Bang, but incorporating a short, early period of exponential cosmic inflation in order to solve the horizon and flatness problems of the standard Big Bang model. Another variation of the inflationary universe is the cyclic model developed by Paul Steinhardt and Neil Turok in 2002 using state-of-the-art M-theory, superstring theory and brane cosmology, which involves an inflationary universe expanding and contracting in cycles.

Multiverse - The Russian-American physicist Andrei Linde developed the inflationary universe idea further in 1983 with his chaotic inflation theory (or eternal inflation), which sees our universe as just one of many "bubbles" that grew as part of a multiverse owing to a vacuum that had not decayed to its ground state. The American physicists Hugh Everett III and Bryce DeWitt had initially developed and popularized their "many worlds" formulation of the multiverse in the 1960s and 1970s. Alternative versions have also been developed where our observable universe is just one tiny organized part of an infinitely big cosmos which is largely in a state of chaos, or where our organized universe is just one temporary episode in an infinite sequence of largely chaotic and unorganized arrangements.

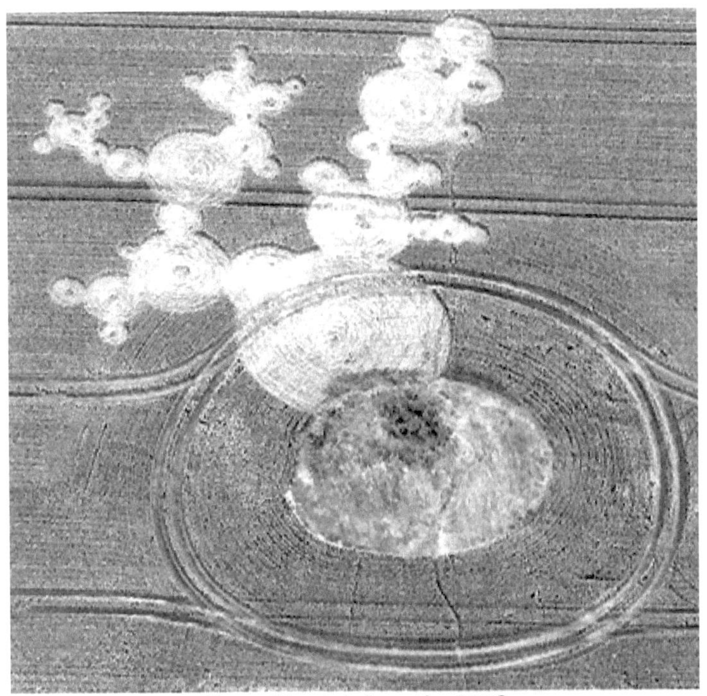

Signs of Fractals from the ONE?

Personal Interpretation of the Relationships

The latter are a collection of aspects of how the body works in relationship to itself, intentional stimuli of our consciousness and general connectivity to the planes or worlds around us. The research and understanding of such systems has helped me to relate to the world and my reality in ever-expanding ways.

Exploring the theories of cosmology, whether reading or listening to those much more educated than I in them, has also given me ways to compare and contrast what I've experienced in relation to them. As an evolving civilization, the need to continue the exploration of our relationships is imperative.

I know there are more things under heaven than our minds can imagine still, yet this at least gets us on the same page, or at least a similar one. I was gradually made aware of the need to consider a holistic view after so many varied contact modality experiences throughout my life, there has to be a holistic view of reality that indeed address the relationships of consciousness, our cosmology and contact with non-human intelligence.

My premise was that if we are ONE, then how does that appear to our senses? Each aspect or expression [intelligence] of the contact modalities was a completely subjective experience, even though I was able to be aware enough in them to ask questions about them. In each there was a component of my consciousness that was 'present' as a participant and yet I was also able to observe and report (even if only in my own mind).

Dr. Mishlove commented that he was impressed with my ability to observe and report on the subjective experiences, which added a dynamic to my writing that many experiencers seem to lack. We are challenged to move beyond the emotion of the experience and question it as an outside observer with an existing knowledge base. The real opportunities come, in my opinion, when we suspend our belief systems and knowledge base to ask questions the take us into new territory.

In a moment of contemplation regarding writing this chapter and being able to offer something unique, my mind found its way to reflecting on a few authors from the late 19th and early 20th century. Helena Blavatsky described Theosophy as "the synthesis of science, religion and philosophy", proclaiming that it was reviving an "Ancient Wisdom" which underlay all the world's religions.

In the 1920s Guy Ballard wrote of Ascended Masters and an 'I Am Presence' that unified human consciousness with a higher realm of oneness accessible through his training with one known as Saint Germain. They first met on the side of Mount Shasta, a common place for CE-5 activity today.

Alice Bailey's vision of a unified society included a global "spirit of religion" different from traditional religious forms and including the concept of the Age of Aquarius. Baird Spaulding wrote of his discoveries on a scientific expedition to Southeast Asia in *Life and Teachings of the Masters of the Far East*.

These early authors who explored the unknown with some voracity, yet only addressed semi-terrestrial circumstances. It wasn't until much later that the UFO phenomenon was introduced in the mid-1940s. Now, if you were able to *grok* (from Heinlein's Stranger in a Strange Land) all that, I'll elaborate more of what I think may be possible in answering the final question. I'll attempt to supply an adhoc unifying theory of everything that may hold some advantages and perhaps correlative factors for the reader.

Trapped in a Paradigm

We're a world in fear right now, perhaps rightfully so with the trends of conditions we feel helpless to change. Is it by design or negligence in our own development? Is there a cross-over in fields of study, a synergy of discoveries and knowledge that we've missed somehow? The author's opinion is there is indeed a virtue in contact that can impact our future in positive ways.

Ufology has a convoluted past and severely limited present with the kinds of information that keeps us from moving beyond the separative notions that seem to keep us afraid, angry, ignorant and immobile in unifying our efforts toward a virtuous path. For some, it is obvious that this virtuous path does indeed exist and has been either ignored or purposefully obfuscated by those who are still in the command and control paradigm.

So many 'experts' are caught up in conspiracy and phenomena, leading us away from a concerted effort to become more self-aware and fearless toward our future. The concerns for our planet and people are real yet many, if not most, feel powerless to change our current course. "Be The Change," still echoes in the memes and themes across the Internet.

Being the Change

How do we 'be the change' in the face of the relentless onslaught of misinformation and intentional distraction from a collaborative and cooperative movement toward harmony among people and planet?

Too many feel the 'harmony among people and planet' is an incredulous notion we will never be able to manifest as human beings on Earth. To the contrary, there have been statements and inspired presentations of a 'new age of enlightenment' for decades.

Ufology has been an environment where money talks and bullshit walks in step with the would-be soothsayers who seem to just want to argue and discredit those who have had real engagement for most of their lives.

Countless folks have attempted to enter the fray with open hearts and genuine intention of serving a greater cause, which includes becoming part of a universal order amongst the stars as we evolve into a planetary civilization worthy of such.

In a fairly recent interview, the Sociology of Ufology, we discussed the history of Ufology and the path it has taken via the promoters and presenters within the community. The same themes are regurgitated with different personalities over time, nearly all of which leave a skewed view of the reality many face.

It is often, but not always, steeped in fear of the unknown rather than embracing opportunities to learn more about ourselves and the universe.

The Virtue of Contact - New Discoveries

In recent years the discovery of consciousness being an ever-growing aspect of 'contact' for experiencers led to the exploration of what 'being awake and aware of our surroundings' really means on much subtler levels.

After all, the subtle clues within synchronicities and the growth in self-awareness is tantamount to the advancement of our understanding within contact scenarios. What were once perceived as threats (confronting our fears) have shifted in experiencers' understanding toward the fearless nature of the higher vibratory realms our 'contacts' present.

Through the tireless efforts of many who see this trend and

want to inspire others to consider a new living awareness, progress is being made. Every question has an answer, though we may not understand it immediately. Are we asking the right ones that lead us to the virtue of contact? Instead of looking for validation of our fears, perhaps we can look for evidence of a holistic system that includes our own advancement in consciousness. In the author's opinion, making sense common is our goal and can be linked to frequency; something that resonates in us all.

Making sense common, you might ask? Across the electromagnetic spectrum and within the thoughtmosphere (collective intelligence) truth has a resonance within our bodies and minds. Indigenous wisdom says there are three brains; the gut, heart and head.

The first brain, which we rarely acknowledge as being the most important, is the gut - the solar plexus. It is where we are connected to everything vibrationally - we FEEL first.

The interpretation of those feelings or sensations is where we get bogged down because we don't have many who've been able to give us clear advice, or at least we haven't looked for it. In fact, though, we've been given clear advice through millennia by those who've risked their lives to share the wisdom and path to inner or self-awareness. The evidence can be found in countless volumes of literature on spirituality and self-development.

It could be said that a good 'gut check' is a primary functional capacity we have for discernment. It could also be said that what we look for intentionally, we always find. The latter serves for both fearful and fearless creations in our lives. Quantum physics even suggests that our realities are malleable until we observe our thoughts manifesting.

What kinds of thoughts are present in your mind?

Channeling has been a 'go to' form of information gathering for many as well. The author has had serious questions about the validity of such information coming out of the mouths of those who profess to 'channel.' Some years ago he experienced a public session with Steven and April White who channeled Ashtar and Athena.

The event left him intensely curious about how connections at that level are made, after being singled out in the audience with their stares during the sessions.

There are many intentional perspectives to consider, yet one stands above them all - the intention to be of service to others. It's a pure notion, free of ego and disturbance yet nearly always the channeler does not remember anything of what was said.

The question of accountability notwithstanding, the information is nearly always pertinent to someone or ones in the audience. Remembering isn't important. It is not a remedial reading class for regurgitation on demand.

What the author found out some years later during his own experience of 'letting go' and allowing a trusted source to engage his body, mind and vocal chords is that the consciousness of the individual is still present, just not locally.

Consciousness never completely vacates the scene, so to speak, it just goes to a different location, non-linear and non-local, where yet another private conversation or experience takes place. It would seem this is a similar experience to many experiencer recounts of either bi-location or teleportation into other realms for the duration of the 'contact' experience.

The non-linear and non-local event simply cannot be recorded effectively because the brain/mind has no reference to be able to do so; articulation of such events is nearly impossible at first. Tenure, alas, does allow one to learn how to communicate the experience in lesser emotive ways over time.

The consideration of the virtue of contact within the experiencer community and Ufology as a whole would vastly change the direction of our thinking for the good of all.

Frequency Matters

Many already know of the <u>Kardashev scale</u>, where the position of a civilization is determined by its access and use of energy, from global to solar system to galaxy. I suspect it goes further, but for now let's just consider what the corresponding intelligence and overall consciousness might be that is associated with the ability of the planetary civilization.

We, as humans, tend to see things from very limited perspective because we simply cannot comprehend that kind of advanced consciousness. We often perceive it as god-like, yet it still cannot compare to the intelligence that drives the bus. Could it also have something to do with *frequency* and our understanding of it? Science has proven everything is energy and has an inherent frequency, so it stands to reason we need to explore further.

In consideration of such, how might a Type III civilization communicate with a planetary civilization like ours which is likely to be around a .5 to .7 at present. Taking for granted that interstellar travel and understanding of universal laws has also developed in a Type III, would they see us as needing help in our challenges with planetary resource management and climate changes? Would they try to communicate in helpful ways? How?

We're aware that free or 'zero point' energy exists but are unable to produce any technology capable of utilizing the principles necessary to access Earth's energy to that degree, yet. There is much discussion about it and perhaps soon we'll discover a way to actualize its potential and at least put a dent in the pollution our current energy production creates. As long as the competitive environment exists, it might be a while before we learn how to share let alone collaborate for sustainability.

Since childhood I've had moments, non-linear and non-local, inspiring an intense desire to understand the nature of how they happened and why. It's led to the same curiosity toward humanity and our evolution. The number, frequency and intensity of events caused some serious challenges; a juxtaposition of realities that facilitated the development of a unique observational quality and capacity for questions of self and others as I grew up. When The Celestine Prophecy came out I felt a nice corroboration with what I knew already.

It was apparent that I could not have open conversations to explore the experiences in my youth and for most of my adult life. Either I couldn't articulate things simply enough, my audience just wasn't able to listen or concerned parental units were afraid of encouraging me.

So I began a kind of ongoing compare and contrast essay in my head. If/then inquiries developed from there, often helped by an invisible companion who sounded like a wise elder in my head. I think we all have that voice, I'm just not afraid to admit it and suffer fools if necessary.

Much later I found the three main questions that made people most curious. *Who is God? Is there life after death? Are we alone in the universe?* I explored them, too, and in so doing my life experiences provided the answers, though they rarely matched with popular beliefs. I certainly didn't expect them, yet I was set up from birth to become curious. I took the bait; hook, line and sinker. It led down an immense and intense rabbit hole.

Leaping from our framework of understanding, limited as it is, into considering the framework of understanding that a Type III civilization would have truly stretches the mind and no-doubt engages what we would consider science fiction now.

Experience Beyond the Norm

Prompted by the knowledge that I was adopted, I had an insatiable curiosity about extended family beyond the physical; a child's curiosity about mother and father God and beyond. I became aware of a companion, an inner voice of an adult male, that I could converse. In the mid-60s answers to the quest turned into nightly trips to orange cigar-shaped clouds, though never able to remember inside. I had this sense of family that I couldn't wait to experience again. The experiences would always start with a high-pitched tone, a frequency in my head that seemed to tune me in, sometimes producing OBEs as well.

During my first year in college, technically a 6th quarter sophomore as a result of the CLEP tests, I continued my quest to know truth. I'd practiced meditation, had multiple OBEs and taken a few trips by then. The internal question continued; the voice more instructional and during an afternoon meditation it asked if I was willing to die for what I believed in. It was synchronistic, of course, as I had a protracted prayer request a week before, to know truth and willing to die for it if necessary.

I chose 'cosmic consciousness' and simply 'let go' of the physical at a perfectly timed moment, which led to and OBE and finding myself in the 'light' described by Sanskrit as samadhi, a state of oneness and complete absorption beyond the mind-self.

I went beyond that as well, to a state of being surrounded by points of light, something alluded to in the Bush era yet quite different in reality and has no lexicon for demarcation. The awareness of this frequency pervaded the experience, yet added to it instead of distracting my attention.

Coming back into this world, even though I was only gone for a couple of minutes, was like entering my body/mind for the first time. It was almost foreign at first, yet delicious as I felt the sensation of the intimate complexity of my body again. The experience was so intense that I couldn't keep it to myself and yet was completely misunderstood in trying to articulate it to others afterward. Ever felt like that?

Looking back it is probably like an intelligence or being from a Type III civilization attempting to communicate with a Type I or less, having no way for the lesser to cognate. It would seem that a Type III civilization, thousands or millions of years ahead of us, would naturally want to communicate with lesser worlds. This is intrinsic among life forms of any strain.

Trials and Tribulations

In spite of the trials and tribulations the journey afterward brought, I was still curious about 'God' and how things worked in the span of consciousness I had experienced, from form to formless and back again. I read every spiritual text I could get my hands on for over a decade.

I got married and began raising a family with four children until I couldn't. It was like a timed release, a divorce I didn't want from a destructive relationship no one deserved. I found myself post-divorce and deeply conflicted at thirty not-so-onederful with only my heart and soul to follow.

I prayed and searched for self again, only this time real people showed up. I was drawn to a group who practiced the Awareness techniques developed by William Swygard in the 1950s. They also followed the teachings of Ray Standford as members of the Association for the Unfoldment in Man. I moved into the house that was their regular meeting place.

The Swygard techniques, possibly the first remote-viewing procedures, were facilitated processes that allowed an experiencer to move about inner planes and explore spirit guides, chakras and Akashic records (Multi-Level) and, in another, traverse 9 planes of consciousness while pausing in each for reflection and understanding (Multi-plane); in essence non-linear and non-local experiential events. I was already prone to easy exits, so the techniques enabled me to explore easily and report back. The job of the facilitator was not just to guide the experience; they recorded the responses to questions along the way.

The experiencer was always in charge of their direction, though, unlike hypnotic suggestions that I found present in many of the regressionists or abduction- related hypnotherapists that I met later, embedded in the Ufology field and not necessarily helpful to it. The absence of fear in the facilitations using the Swygard method was profound. A decade and a half later I created a digital version of the Multi-plane that wasn't as personal, yet offered a wider experiential playing field for others.

The events prepared me for another spontaneous event, an answer to questions about God and the Trinity, not necessarily from a Christian perspective either. In preparing for a Multi-level, I was invited to join an entity I met in college that is what one might call a 'guide' or 'ally' in Castaneda's world. He escorted me across the universe in what seemed like a kind of wormhole, conversing about my 'progress' as we went, and arrived at a display that instantly filled me with awe - three huge white gold-tinged spheres surrounded by perhaps a dozen much smaller bright green spheres.

When I went to see **Contact** for the first time nearly a decade later I wept during the scene where Ellie went through the wormhole and arrived at the peripheral view of a magnificent starscape. The display seemed like someone had re- created my own experience and put it on the screen for others to share if they could. It was so impactful I didn't have the desire to talk for hours afterward. I had the same feeling after my only experience of channeling during the same period of engaging Swygard's work. The sense of awe made me speechless.

It wasn't just the scene of the spheres that gave me pause. There was another voice, like several as one, that seemed to come from the spheres in answer to my question about God. "We are not only your forefathers, we are also the forefathers of your solar system." So where does one put that in the collection and referencing of files? What it did was give me more questions, the primary being, 'if this is the macro, what is the micro?' I still didn't comprehend 'what' the spheres were, though I knew they were the Creators as we know them.

If this experience was available to me as a silly human with questions, what might a Type III know about the structure of creation, the intelligence that permeates it and the capacity to interact on a personal and planetary basis?

Frequency of Information

Several years later, in between viewings of the newly released 'What the Bleep,' and a discussion with William Arntz, it occurred to me that it must be the electron, proton and neutron. The 'space' is where the consciousness or intelligence that manages manifestation resides. I realized nearly two decades earlier that the 666 was the carbon atom, the science behind the number of knowledge and wisdom, the number of man.

The bridge between science and spirituality was making sense yet I couldn't even begin to grok the equations that would define it mathematically, though I knew they had to be developed or already exist somewhere. Like the primer in Contact, there had to be something that explained the experience in mathematical and/or scientific terms. In the background in every instance was this frequency, subtle and discrete, yet it was there. I think in some respects it also appears as a cognitive relationship; conversations that matter in our head.

The frequency was perplexing, though. Although it vacillated to some degree, it was always present in the experiences and, the more I was able to listen it seemed to always be there in the background. Sometimes it would get louder for a moment, be in one ear or the other more prominently.

The standard response was that it was Tinnitus, though no medical tests were able to prove it. It became a constant companion when I was young and is still with me today.

Does focused listening promote or provoke experiences?

Enter the extended communication and events with non-human intelligence and the appearance of various forms we now call greys, insectoids, mantids and reptilians along with some humanoid types. I didn't put any star-system references to them, nor did I ask. It didn't seem necessary to know. The frequency, though, was much more intense. I hadn't made the connection with the Solfeggio frequencies yet, but I did notice subtle differences with each race.

I was visited multiple times over a several year period in the early 90s, one rather strange that was a little embarrassing. I felt a certain familiarity with all seeming to know me beyond any circumstantial events, like we'd been in communication for some time. Each had something to offer to tie-in the previous knowledge I'd gained from childhood through the Swygard period in the summer of 1989, as well as further initiations into what was being referred to as the Ashtar Command. A startling event came a decade later.

The initiations and interactions were nearly too much to take in and still have some functionality in the world, as in holding a job and paying bills. I managed to do it though, as well as produce and host a television show to explore the inner promptings of others and how they dealt with overcoming fears in their personal lives and professional careers. I interviewed over 200, from contactees to city managers to ethnic leaders and university provosts.

These conversations provided a good anchor to explore the dynamics of what had been shared through the previous encounters, from college to the present at that time. I'd always been more alert with synchronistic events, yet what I noticed was events and meeting people began to 'just happen' on a more regular basis, apparently as a result of my contemplation and questions toward integrating the understanding I'd been given in pragmatic ways.

Giving a glimpse of how people worked through various trials and tribulations to our audience might help them feel like they aren't alone in the plights.

I was led to various people during this period as well and was instrumental in organizing events that included metaphysical as well as extraterrestrial perspectives and had from 150 to over 5,000 patrons, earning a MBA and Secondary Teaching certification.

I had a conversation with Dr. Edgar Mitchell during one of them, sharing our interest in consciousness, frequency and contact with intelligent beings beyond our scope of understanding. He shared the understanding that frequency, what was in the silence within, had much to do with advanced intelligence. He launched the Institute for Noetic Sciences as one of the results of his trip.

I attended many events, had conversations with authors, 'experts' and patrons. What I observed with rare exception was a fascination with sensationalism, conspiracy theories and cover-ups that were far from the reality I had been experiencing. The conversations with presenters seemed strained when I asked questions about direct-experience, or understanding garnered, regarding anything to do with consciousness and/or the engagement of other races toward some beneficial end. None seemed to want to talk about inner experiences, let alone the frequency many experiencers reported. Perhaps it just wasn't something others experienced.

Then things got really intense for a couple of years, a more public engagement of Ufology to the extent of accepting a leadership role during a conference in Camp Verde, Arizona. My life was turned upside down, or right side up depending on how you look at it. I thought I was prepared, yet a couple of months later I had a week-long symposium onboard a ship and was emceeing the entire thing.

I built my first website with articles and information from scouring the web, including all the resources I could find at the time and soon was getting over 25,000 visits a month. Then I got disillusioned with people and process and took it down, redesigned it and moved forward. I suppose I left the sandbox for a while.

I got involved with other passions for a while. I played drums for a few outcasts and social misfits, taught high school, ran the electronic scoring system for fund- raising go-cart races around the country on weekends. I went from public to charter high schools and earned a Master of Arts in Organizational Management in which I wrote a business plan for a peer-community school/village.

Then I began facilitating pre-construction partnering workshops, team-building for construction teams that had budgets of over $20 million. As an example, the F-35 Operations and Maintenance Facility at Luke AFB was one of them. I earned a transformational life coaching certification at about the same time I started facilitating and had two businesses brewing.

I took a year off and traveled the country with theater groups and a couple of musical legends, worked for a performing arts center and eventually took over a local UFO Meetup group in 2008 (now defunct). I got out of the ufology stuff for a few years and then had an opportunity to create UfologyPRSS.com, a multi-device personalized RSS (PRSS) reader. It is designed to keep you up-to-date with all the latest news on Ufology by instantly connecting you to the best UFO news sites, books, blogs, podcasts, videos and more.

The two businesses became three when I launched Planck Social Media Agency in 2016 and was able to update another website I own/manage, IndependenceArizona.com, to make it easy to get the Independent Party registered by the State of Arizona. The former is doing alright, the latter... not much interest.

A couple of months into the 'event' of 2020, I joined forces with a friend to create WeCanDoBetterArizona.org and solve the crisis facing kids aging out of foster care in Arizona. If we can do that, then we'll have the tools to solve other problems and teach others how.

Frequency within UFOlogy

In 2010, the International UFO Congress moved from Laughlin, Nevada to Fountain Hills, Arizona and I got a few members of our Meetup group to share a vendor table. I'd published a few books by then, too. The following May I had another visitation, only it was different this time. I was awakened by some familiar folks, tall greys I'd known for decade and a half.

They showed up, announced I was getting an 'upgrade' and proceeded during our brief conversation. I was actually thrilled to feel and hear them again. The 'upgrade' felt like something was installed in the back of my head. The sensation was rather intense, but certainly not painful in any respect.

The following year, during the IUFOC, one of our group told me she was shown/told of a device that apparently plugged me into the cosmic computer that was in the back of my head. I hadn't shared my 'installation' experience with anyone except the woman I was living with at the time... no one. It was interesting for me as back in 2000 I had a conversation with Dan Winter, where without provocation he shared with me that he believed ASHTAR was the acronym for the cosmic computer. Go figure...

I continued with the IUFOC for a few years, met the folks at a business expo who designed a digital wire-frame template for curating RSS feeds on virtually any topic. We soon created UfologyPRSS by early 2013 and launched it at the IUFOC that year. The owner of the company had an interest in UFOs so the timing was perfect, yet again. I also did an interview in the foothills of the Superstition Mountains the same month in the backyard of Cynthia Crawford, the et sculptor (RIP).

The interview in the desert was a public leap into the rabbit hole, a transparency in conversation I hadn't been able to experience to date. I have to give credit to Adam Abraham for being centered, open and asking the right questions without prejudice of any kind. I exposed myself completely, as transparent and vulnerable as I've ever been, yet felt comfortable with Adam and the surroundings in doing so. I noticed the fluctuation of the frequency during it, felt an openness of information flow and yet completely conscious of participant, observer and source.

Helpful Events

Over the next few years, communication of intention and supporting information regarding the evolution of our planetary civilization became more prevalent in all my 'contact' events. It had been obvious that our leadership, around the world, had been steering us toward destruction and needed some serious redirection. I'm just one guy, so how could that possibly work? I was reminded of the thoughtmosphere, the points of light and the volunteers who've come to serve.

I noticed a shift in the kinds of movies that were addressing the extraterrestrial element. They went from the standard Hollywood model of fear-based non- sense, like they'd done with Travis Walton's story, to something more poignant. Film makers were actually looking at the non-human higher intelligence and helpful qualities beyond our fear-based responses to some perceived alien threat. At the same time, the conversations in Ufology began to consider a more benevolent side of the paradoxical equation they'd been running for decades.

Consciousness, the intimate connection with everything, began appearing in open discussions. People began talking about hearing frequencies that accompanied events and the notion that there really is no threat from space. Along the way I created several websites to demonstrate alternatives and share with others who felt connected to the same ideals. These included my original website remodeled, Mothership Cafe, Planetary Citizens, Independence Arizona and UfologyPRSS.

Back to the frequency… It's not just the scheduling of events, but the actual frequency that many report being able to hear either just prior to or during contact events. What is it all about? How does it fit with experiencer explorations into realms we barely know anything about, except the references in some ancient spiritual texts?

Could giving it more attention enhance our ability to explore knowledge and understanding of this cosmic order we're discovering in our quest for truth?

Science of Frequency

Let's go to the science side of things. The electromagnetic spectrum covers a wide range of frequencies, waves that may become particles in the awareness of observation that spans creation. We have brain waves that are within this spectrum, demarcated into ranges of consciousness from waking to deep trance. We can measure them, but we don't really have the ability to hear them in our 'normal' range of perception for the most part.

Some time ago, in the late 80s, I was introduced to the Shabda as the vibration of all things, like the combination of spiritual and material substances on Earth, also known as the 'sound current.' Since that time I've notice more explanations of it, yet it is essentially the same, only expanded to the audible stream of energy, emanating from the center of creation. It doesn't take a great leap to connect the electromagnetic spectrum to this audible energy.

Within that frequency, when one can focus attention, are many realms that open to experiential moments. I can't say that I've consistently visited them, though in what appears to be serendipitous and synchronistic moments I've had access, sometimes for fleeting moments and sometimes for extended periods. I mention the ability I believe we all have to enter those realms in a conversation with Dr. Mishlove, where we discuss a 'new world order' that is basically an acknowledgement and alignment with these 'harmonic' frequencies that permeate our reality.

Something within us connects us to everything else, though we may not understand it or, more appropriately, be able to utilize it to evolve into a greater experience of being. The evidence of the possibility to do so is contained in every spiritual path, yet there has been no science that brings the hemispheres of the brain together to work as one. The analytical and creative have essentially been at odds for centuries if not millennia.

How might we bridge them?

Tom Campbell, who began with Bob Monroe in exploring OBEs to develop both material and practice methods for adventures, states that his work has been an effort to explain the right-brained worlds to left-brain thinkers. Could the corpus collasum, the membrane between the two, be like the center line of the Tao in defining a permeable boundary between the two?

In considering the primer from Contact and how it brought a two-dimensional image that made no sense into a three-dimensional view that allowed the building of the machine, also contained in a frequency that was deciphered with math, is there something we're missing with the Tao?

This may be a stretch for many, to consider a three-dimensional model of the Tao and what it might reveal. We know that energy pulses through our neurocircuitry to convey messages. Light pulses also travel the glass tubes to deliver data packets across the web.

If we look at the Tao, the yin/yang symbol, can we envision the pulse of light traveling the strands of DNA in the double-helix, one side light the other dark and reversing as part of a data delivery system deeply embedded in our bodies as a mechanism for consciousness to emerge and align us with a higher order?

Speaking of DNA and hydrogen, there is biology and science that bring the two together in a most elegant way as well. We already know that our sun is comprised of 75% hydrogen. It supplies us with energy for sustenance on so many levels. Harnessing the energy of our solar system, including the sun, puts us in the Type II Civilization ranking. Let's look internally.

The hydrogen atom is the bonding agent for the helix of DNA. The complementary base pairs of guanine with cytosine and adenine with thymine connect to one another using hydrogen bonds. These hydrogen bonds between complementary nucleotides are what keeps the two strands of a DNA helix together. Fascinating, huh? Hydrogen atoms, like fractals, are everywhere. Let's dive a little deeper and hopefully it will make sense.

Frequency of Hydrogen et al

The frequency of hydrogen is roughly 1420 MHz and sounds like radio static. It's interesting to note that the voice of Vrillion, a radio transmission in 1977 that interrupted a British newscast, has a static within it that is consistent throughout the broadcast. I don't know that any analysis has been done on the static, yet the message from the Ashtar Command has mixed reviews.

The same message, essentially, was contained as binary code in a crop circle in August of 2002, "Beware the bearers of false gifts and their broken promises. Much pain but still time. Believe. There is good out there. We oppose deception." Sounds like a great message with some concern for us, right?

More recently, as a result of reconnecting with a television show host who inspired me as I was doing One World in the early 90s, I was introduced to a fascinating model of reality called the <u>Triadic Dimensional Distinction Vortical Paradigm</u>, TDVP for short. Developed by Drs. Vernon Neppe and Edward Close the mathematical model presents the concept of consciousness, space and time being tethered across a 9 dimensional framework.

Reality involves a unified wholeness of the infinite and finite with the infinite pervading the finite experience of content and extent in the Space, Time and "Consciousness" Substrates. This finite-infinite reality is relative to all dimensional-distinction factors, and experienced subjectively (by individual units), with perhaps a collective or shared experience available. We've had some of those during group meditations I've facilitated and participated.

A recent movie titled simply, **UFO**, and stars one of our favorites from the X- Files, Gillian Anderson, is a story about a sighting and the discovery of a frequency contained in static in the 1420 MHz range that is recorded by the tower at Cincinnati airport.

A young college student, a gifted mathematician, discovers it contains an elegant display of mathematical precision that has relevance in current applications of known equations and formulas used to explore communication and locations in space.

These equations revealed things like the frequency of the basic hydrogen atom, the most prevalent atom in the universe. It also introduces the <u>Fine-Structure Constant</u>, denoted by α, which might have some relevance in referencing spiritual texts as well.

As I was writing this it occurred to me that the phrase, "I am the Alpha and Omega," might have some scientific significance, energy and matter within us all. The anthropic principle is a controversial argument of why the fine-structure constant has the value it does: stable matter, and therefore life and intelligent beings, could not exist if its value were much different.

The symbol Ω has applications throughout mathematics and physics in surprising relationships. Could it be that the phrase actually integrates energy and matter instead of alluding to some personage in Revelations? Could it possibly mean we are all able to access this awareness or knowledge?

With all that in mind, the intention is to create a flow, a stream of consciousness that assists in investigating the current presentation of information across a wide-spectrum of experiencer perturbations, questions that no doubt are surfacing in the consideration of a new living awareness that is permeating the consciousness of many today.

We do have to think differently and pull ourselves out of the deceptive and divisive notions of traditional fear-based human perspectives. We also need to withhold judgement of others sometimes, especially when new information or understanding might emerge.

We seek flow; the unimpeded data transfer and development of a new kind of planetary civilization that allows us earthlings the ability to join a greater universal experience. It is my opinion that as we increase the energy of what we call love toward each other, even focusing on and intentionally sharing the 'frequency' from within, our world will change for the better rapidly.

As it is, we don't really know if we have time, yet time is all we have. Let's use it wisely. There is, however, a notion that time may be just a measurement of the change of entropy. Holistic systems tend to gravitate the opposite direction, toward harmony, so perhaps the time frame is actually extended.

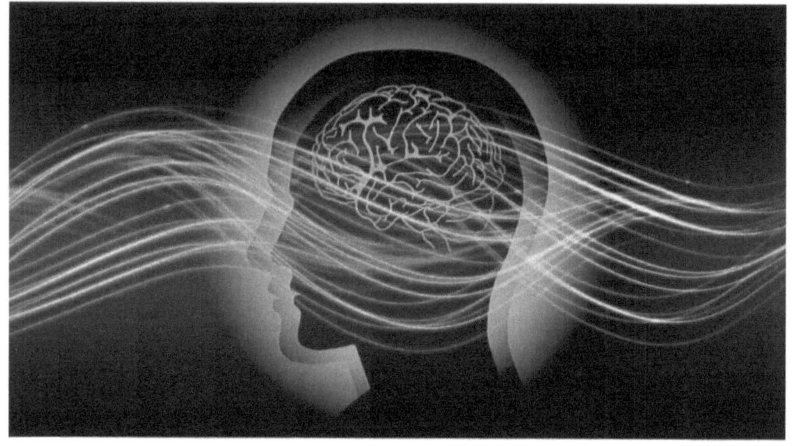

How Do We Qualify Ufology or Any Information?

How do we 'vet' the information and people providing it so that the quality is there; so that sense is made common?

It would seem that a consistent method for discerning or discovering truth, even if relative, would be available to the intentional seeker. Of course there are varying degrees of the depth of the quest for truth, from cursory to quantum, yet consistency is key. If we accept that we have built-in bullshit meters, then we can at least begin to discover them within our awareness.

What kind of tell-tale signs does 'truth' have in this, or any, environment? We pay too much attention to paper and not people, in my humble opinion. People will show themselves to those who know how to look. When they do, believe them. Not necessarily what they say, but who they are and how they show up authentically. You don't have to be an expert at micro-expressions. Truth has a ring to it, literally, though it's an internal ring.

Perhaps truth, vibrationally speaking, is like a covalent bond where electrons are shared in a kind of quantum entanglement. The *resonance* of truth occurs in the receiver as well as the transmitter whether person, place or thing. Our bodies are both, so it would make sense that the reciprocal resonance is at least felt if not understood. We aren't accustomed to relying on feelings or sensations, though. That's where things get a little messy for a time while we explore and practice honing our skills.

There are certain characteristics we can count on, some external tests we can perform initially as we deploy our discernment tools. Until the inner tools are developed, the outer ones are most helpful as one is able to observe without attachment. Like the progress toward self-awareness, we must observe our intentions and distractions. Here's some considerations from the works of <u>Brent Cunningham</u> on using the external... makes sense common.

Popular Reasons Why People Believe

Before we take a look at what I think are the three essential tests for evaluating any truth-claims, let's briefly look at some popular, but less than stellar reasons often used by people to test whether or not something is true—the claim and then a response:

*1. **Instinct** — "It seems true to me."*
To this statement we must simply ask . . .
. . . Why? What is guiding one's instinctual ability to judge between competing truth-claims?
*2. **Feelings** — "I like the feeling I get." or "It feels right."*
. . . Can feelings be mistaken?
*3. **Wish fulfillment** — "The God that I (want to) believe in would never*

send someone to hell."
. . . Maybe not, but the God one wants to believe in may not be the God who actually exists. We can never determine objective truth simply by what we want to be true.
*4. **Custom** — "This is the way I / my family / my culture has always been."*
. . . Can they be mistaken? What if one came from a Nazi background?
*5. **Popularity contest** — "Well, 'everyone' believes it."*
. . . Again, can 'everyone' be wrong? Counting noses is never a good means of determining truth.
*6. **Pragmatism** — "It works for me, therefore it must be true."*
. . . However, almost anything can 'work' for a time. Truth works overall. For instance, a belief may 'work' at ridding oneself of guilt by denying it rather than absolving it.
*7. **Significance** — "It gives my life meaning."*

. . . Is the meaning simply the result of wish-fulfillment or is it connected with reality?

Obviously, all of the above reasons for believing something to be true seem deficient and not the best tests to determine whether or not a truth-claim is really true (although, #s 6 & 7 do seem to be better candidates that #s 1-5). Nevertheless, I think there are three criteria for testing truth-claims which, when used together, offer us the best chances of determining true beliefs. Here they are.

*1. **Internal Coherence** — This is a test for rational consistency in ufology information. This asks if a belief makes sense? We need to determine whether beliefs are rationally consistent within themselves and in relation to others' beliefs of one's larger worldview. Some beliefs are known to be self-referentially inconsistent, or self-defeating. An example might be the belief that all "knowledge" is scientific knowledge. This is obviously self-defeating because the belief itself is not a scientific statement. Therefore, the belief dies by its own standard for knowledge.*

Another way a belief can fail this test is when two beliefs are in contradiction with each other, meaning that at least one of the beliefs must be false. So, we must ask if the beliefs hold together? For example, if a person is a Naturalist (believing that human life is accidental, random, and without ultimate purpose), he cannot then introduce the belief that we have an objective moral obligation to treat another person justly or with kindness.

*2. **External Correspondence** — This test asks if a belief fits the facts of reality in ufology information. Does it correspond to the real*
world? Proposed truth-claims must have explanatory power, or the ability to give account for our experience of the world (whether it be history, science, psychology, human nature, etc.).

For instance, a worldview can be tested by its ability to explain cosmological questions like the origin of the universe; anthropological questions like the existence of minds and free will; moral questions like the existence of evil and our experience with guilt.

Further, the explanatory power of truth-claims should (a) be comprehensive in scope—able to explain more or better than alternative theories; (b) have predictive power—suggesting new evidence and problems; (c) have precision—accounting for more details;
(d) be illuminating—integrating otherwise unrelated data; (e) avoid ad hoc hypotheses—functioning only to explain away counter-evidence; (f) be simple—not needlessly multiplying the basic concepts, assumptions, and principles of an explanation.

*3. **Functional Adequacy** — This tests the livability of a truth-claim as a belief in ufology information. Is it a viable belief "on the street"? Does it work in real life? Some views sound good on paper, but are proven false in the laboratory of life. Consider an eastern guru who asserts that the physical world is an illusion, yet he still looks both ways before crossing the street.*

A person cannot live out such an illusory belief of the world for very long
(or he'll be hit by a bus before long!). Even more than that, a belief system must integrate one's life. It must incorporate and meet the deepest human needs.

SUMMARY

*In evaluating the truth or falsity of propositions/truth-claims/beliefs of ufology information, we must be sure to always look for three things: internal coherence (the **logical**), external correspondence (the **factual**), and functional adequacy (the **livable**). For a belief to be true it must be meaningful, it must line up with the real world, and it must not only help us survive in daily life, but allow us to flourish. Consequently, these are also the three areas in which our thinking can and does go wrong:*

logic, facts, and values.

Applied Understanding

In the field of Ufology there is so much information that has
no solid proof, no tangible substance, other than the ability to
string what little facts we have into some kind of logical
presentation. We all wish this was different, yet beyond the
little credible and tangible material we have, much is still in
the subjective realms of experiencers who are often bereft of
the cognitive skills to manage, let alone interpret, their
'contact' events.

Truth is discoverable, of course, yet we have to run it through
all the tests for coherence, correspondence and adequacy. It
has a resonance to it that meets all those standards, a
measurable sensation within our array of receptors. We have
well-developed receptors already categorized and labeled as
the 'clair- alls'* of extra-sensory perception. We all have them,
every one of us, but few actually use them, probably because
of the 'I want to believe' factor within Ufology.

Consciousness - being awake and aware of our surroundings,
not just sentience, is the prevailing discovery of the time.
Consciousness, in the sense of a connected awareness and
experience of a greater theme in our lives, is an experiential
reality that bridges quantum (discrete) with the gross (whole).

The resonance of both cohesion and coherence is what we can
sense with the finer senses within our awareness. It is the
challenge of accessing and practicing these senses that is our
current challenge as we grow in intelligence and knowledge.

The constructs of reality also have significance, even if we
don't understand the laws or principles involved. For instance,
The Triadic Dimensional-Distinction Vortical Paradigm is a
metaparadigmatic model developed equally by Drs. Vernon
Neppe and Edward Close. In essence, it demonstrates that

consciousness, space and time are tethered together across 9 dimensions.

It would seem there is a singular or individuated consciousness that is distributed across dimensions yet capable of interacting with and observing each perhaps simultaneously.

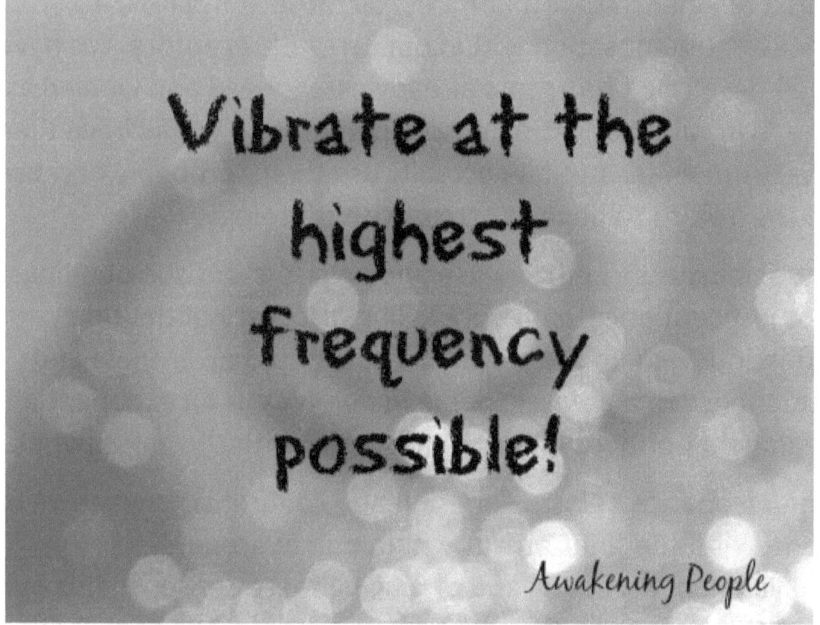

A Possible Unification Theory?

A Little Preface

Before I get into the elucidation of a theory, I'd like to offer another experience that has made a huge impact on me. It began as I was being prepared to venture out, into my experience during what was to be a Multi-Level Awareness process. There is an initial guided process that, in essence, loosens the light body (astral body) and facilitates the use of the 'light body' to travel on the inner planes of consciousness. It is challenging for some, freeing for others.

The purpose of the facilitator was more as a prompter and recorder; to maintain a verbal link with the experiencer so that reflections could be made afterward. In this way there was some record for reviewing for the experiencer. Records are great for review and it further solidifies the experience in memory.

From my training as a hypnotherapist, I find the similarities between the two striking, yet the direct experiences are in the present moment, not a recall of previous events often used for deeper exploration of 'contact' experiences. However, past lives can be visited as a function of the technique. There are multiple choices the experiencer can entertain prior to entering the process.

This was definitely a metaphysical experience, yet it seems to correlate much of the scientific research and referencing of the 'Holographic Universe' of Michael Talbot's discovery and sharing. In fact, this paradigm has been present in several physicists and scientist's exploration and revelatory considerations. I suspect Tom Campbell knows this well as his lectures and book, My BIG T.O.E, does a great job at explaining right-brained activity to a left-brained person while presenting the notion of an avatar in a virtual world.

Among those physicists and scientists are Alain Aspect, David Bohm, Karl Pribham, and Karl Lashley. Complexologists at the Sante Fe Institute have created yet another theory that the Universe actually seeks to replicate itself, based on their study of the math, physics, and science surrounding cosmology, quantum environments, and thermodynamics to name a few.

James Gardner also includes this notion in ***Biocosm: The New Scientific Theory of Evolution***. Striving to understand how our universe works leads us to better understand how the human enigma works in its interaction and living amongst the stars. It was there the equation variable for the appearance of carbon got my attention, being 32 decimal places long. It indicates a very intelligent design with precision activity.

Apparently instantaneous 'faster than light' communication between individuals, as well as brain cells, is a reality and violates the long-held notion that Einstein fostered: *no communication can travel faster than the speed of light*. The information age, along with scientific exploration made possible through discovery and development of new paradigms and technology, brings with it the ability to question everything we have learned about humans to date. Hopefully the logic path leads us to new questions and valuable rewards.

Questions often scare people, especially ones that have obvious answers that have been hidden from view or are unanswerable just yet. This includes being able to cross-reference, if you will, the scientific and spiritual manifestos of our history on planet Earth. So let me take you on a little journey. You don't have to believe it was real, although I will do my best to relate it objectively. You'll love this trip and the sublime awareness it brought.

Changing My Thoughtmospheric Conditions

On this particular afternoon, I was in process of preparing for one of these multi-level awareness journeys. I began by doing some deep breathing relaxation to prepare for the process. We had just started the actual preparation when our process got interrupted.

I was feeling my body and mind relax when all of a sudden, Zephyr, a guide I'd known since my late teens, showed up in my mind's eye and gave me that look of anticipation. You know the one, where you know you are about to have something very profound happen and you let go of anything else on your mind. So it was. He simply motioned with his arm and said, "Come." And so I did.

I exited my body with such ease that I hardly felt the departure. Now the interesting thing is that by practicing this technique, I had the ability to describe what I was seeing and hearing along the way. It made it more visceral for both the facilitator and me. I'm sure he wasn't prepared for what was about to happen, even though the practice of the technique included taking copious notes so the event could be reviewed.

As soon as Zephyr showed up, I was communicating my experience to the facilitator, who was now about to be tested in his ability to respond in a much different way. I'm sure it had to have been pretty weird sitting in the chair next to me that day! I had to explain who Zephyr was later, so let me do so now if I may digress for a moment.

I'd met Zephyr while studying metaphysics with a small group of explorers in college in my late teens. We all read many books and discussed them as well. I was in process of reading Carlos Castaneda's works. I think I was on Journey to Ixtlan at the time. I remember one of the consistent threads of other spiritual works being the existence of spirit guides or allies, as he called them. If these are present, then we should have access somehow.

I had inquired within, through prayer, regarding the existence of and ability to communicate with a guide or ally or guardian angel, if you will, that was 'assigned' to me. During a meditation one afternoon about a week later the name Zephyr and a face of an Indian appeared in my mind instantaneously. I was a bit shocked to say the least, especially the way he just popped in out of nowhere and seemed to present.

He appeared as an ancient Indian, with eyes so deep and simultaneously cold and warm, peering into the depths of my soul, yet with respect of my being. I was able to do some research later, assisted by a friend that was an adept at automatic writing, and found that his incarnation was over 20,000 years ago in what is now the Southwest US.

I think it had a lot to do with my move to Phoenix later. I've had numerous occasions to journey with him since. If you've had similar experiences you can relate. If you haven't, well just consider it a great story. Maybe, just maybe, there is one on the way for you. No doubt your mind has been swirling.

This session was a bit different because there was now a witness to record my experience as I related it. As I exited my body, he became a sphere of light with his face as the only feature on the front of the sphere. I first noticed his profile, with that large Indian nose... rather reminiscent of Jimmy Durante.

I found it a bit humorous and told him so. Nothing can be hidden in those realms as even the fleeting thoughts are as apparent as turning a light on in a darkened room. So, just let it fly. You'll find your way sooner or later. His humor caught me by surprise on many an occasion.

So as I was entertaining myself at his expense he turned and faced me with that look again. I asked him where we were going and he only responded that I would find out shortly. He began asking questions about what I'd learned since the last time we journeyed.

I could see faintly the points of light, elongated, as we were whizzing by at an incredible speed. Then nothing was in my peripheral vision, until I felt us come to a quick stop. It seemed like the 'nothing' lasted for about half the journey or maybe I was just distracted by our conversation along the way.

I suppose the only reference to the travel route would be the speed of thought. According to the Urantia Book, an interesting exploration of the structure of creation, the speed of thought is approximately 841 trillion miles per second.

The speed of light is 186,000 miles per second. In answer to my question of direction, Zephyr told me just to wait and see.

I probed again and his silence was absolute. I then spoke to him of other things that I had experienced lately and he responded with analysis, some sick humor, and engaged me in some great bantering that left me feeling very humble. I had some pride in the depth of understanding I thought I'd been expressing to him - consistent for a student/teacher relationship. He had a way of slicing through my ego and putting me in my place.

I think we probably all feel that way when our bubbles are burst by our elders, bringing us back to a humble reality. One of the aspects of the Path is to be so rooted in one's own knowing as they are expressing it, yet being able instantly to let go of the attachment of its truth in order for further depths of understanding to occur.

Sometimes the knowing shifts, and must be left behind as merely a stone on the path. At other times, the knowing evolves into a much deeper acknowledgment of 'what is' in the conundrum of 'What is reality?' A moment of awe usually follows.

This bantering lasted for about eight minutes as we made our way back, according to the facilitator, as the arrival at our destination brought it to an abrupt close. I suppose that the constant recounting of conversation helped to anchor consistent patterns of thought and holographic reality at that point.

It was very humbling, nonetheless, arriving in awe as well. The discussion also had a profound effect on the facilitator's life as it verified some inklings he'd been entertaining as possible answers to some of his own questions about the nature of creation and reality as we know it. Questions scare people more than the answers sometimes, but it's always good to ask.

I noticed a slowing of momentum and as soon as I did, I felt an abrupt halt. We had arrived at a solar system that was beautiful beyond belief. If you've seen Contact and Jodie Foster's character as she witness the beauty of space you can imagine the feeling of awe that overcame me then as I witnessed a nearly inconceivable scene.

We had stopped outside the perimeter of the largest orbit of some bright green planets that revolved around three huge suns that appeared so white yet shimmered with rainbow sparkles. They appeared to be equidistant in the center of this system with nine planets around them.

Again, the sight of this brought up the feeling of awe that came from such an indescribably beautiful display, much like the movie 'Contact,' only this was nearly a decade before its making. I felt huge in that moment. I actually cried when I first saw the movie, realizing that someone had 'captured' the emotional and psychological sense of the scene when Ellie first exited the wormhole and saw the galaxy in front of her.

I only had a moment to take in the beauty before I heard a voice that felt like the combination of the suns speaking as one. I listened intently. The sensation was like a river of energy flowing through me at an incredible rate of speed, like the first fall on a toboggan run.

In response to the natural curiosity and questions I must have been thinking, the voice (s) said, "We are not only your forefathers; we are also the forefathers of your solar system." How does that work? Why am I being shown this? Really?

In spite of my questions I felt such a deep connection to those words and the resonance of truth that I was thoughtless for a moment. I wish there was a word to describe it. You get the picture... I was blown away. I regained my perspective for a moment and wanted to ask questions.

Of course, right on cue, Zephyr said, "Okay, that's it, time to go." I certainly wasn't ready to go. It took me a moment to regain composure again. I really wanted to stay there and ask questions. Zephyr cut me off immediately saying, "Nope, you've got all you need. You'll figure it out."

I needed more feedback. This seemed like a closed loop with no chance of opening it for now. My mind needed more information to process. The statement was just too simple. I argued for a few moments and then gave in to the trip back, realizing we were already on the way regardless of my desire.

He never did respond to those direct questions... damn stoic Indian. He was kind enough to offer some things to ponder, though, which gave me much more than I realized at the time. At one point I wondered if he was an Archon, but dismissed it because of the shared history.

Reconnoitering and Review

Once back in the room and relating the journey to fill in the gaps of my description of the scenery, what I found interesting is that each leg of the journey was consistent. Even in what was obviously a non-linear and non-local event, time was consistent.

According to the facilitator's watch the journey of going and coming were almost exactly 8 minutes long, which was more than just coincidence. This would indicate that it was indeed a journey that crossed some major distance at the speed of thought or something similar yet to be measured by modern science. The Urantia Book says the speed of thought is 841 trillion miles per second and is the only place I've ever found a reference.

Just how the actual vehicles became 'real' or the method of travel used was something I didn't understand as a repeatable process. I did feel an unusually high resonance with the idea of these *suns* being the foundation of our Trinity. I just made sense somehow, that these three essences would combine in the creation of different beings and the worlds they would inhabit.

I had no idea how the additional questions would be answered or if they would at all. But they were over a decade later after a conversation with William Arntz and in the week following my first viewing of 'What the Bleep Do We Know.'

William was introducing the film across the country at the time and had stopped in Tempe, Arizona for a two-week engagement at the Valley Art Theater.

Just imagine what this could mean to the developmental theories of creation and our scientific explorations into the depths of material structure with the proton, neutron and electron. Something created and facilitated their movement into the elements, maybe more.

I was again thrown into matrix of experience and thought-provoking notions about the nature of human and solar system interaction, not trying to create a model with my intellect, letting a holistic model emerge on its own by asking questions and observing how the answers came. I can only describe as a series of non-linear and non-local moments of salient points being presented through imagery and accompanying thoughts.

Could it be possible that the proton, electron and neutron are the micro- reflection of a macro consciousness?

I do know this: it was much more that my feeble brain was able to interpret, yet the imagery, voice, and travel were very real. I'd been on many journeys to date and this was one of the most profound and potentially rewarding in the development of a cognitive model that removes conjecture and superstition regarding the nature of how we got here, or at least of our foundation. Would you agree? Perhaps not, yet the notion is intriguing to say the least.

Further Inquiry and Patience

To understand the mechanisms of creation from that point of view requires some intense study of the internal structure of creation. We don't really have the tools to do that effectively, yet. Metaphysics seeks to explain this in my opinion, but not completely.

We still don't know just how we learned to condense into form. Quantum physics and complexologists are moving closer to explaining the next levels of garnering consciousness in mathematical terms that result in patterns encountered in inner space. Quantum entanglement is the new buzz word in the metaphysical realms as well.

Many years later, after looking into M Theory and String Theory and weighing their potential of experience (theory needs to resolve to experience), I came across another model that assumes that there is a fundamental tethering across consciousness, space and time. Vernon Neppe and Edward Close developed the new model - Triadic Dimensional Distinction Vorticular Paradigm (TDVP).

Triadic Dimensional Distinction Vorticular Paradigm (abstract of paper)

> *Despite our current reductionistic materialist paradigm allowing explanations of almost all aspects of our experience, certain areas of scientific endeavor are still contradicted. This motivates the need for a new paradigmatic approach to integrate several different scientific disciplines and to postulate a new and comprehensive model, producing a paradigm shift. But this is not easy and has never before been achieved: Previous models have often ignored the fundamental role of an extended consciousness. Even in the score of Theories of Everything (TOEs) recognizing consciousness, few include multiple extra dimensions, and only in our Triadic Dimensional-Distinction Paradigm (TDVP) are infinity, order and life fundamentally incorporated.*

> *TDVP regards the unification of space and time as insufficient, and postulates that space-time and a broader extended "consciousness" (STC) are fundamentally tethered from their finite origin, such that even when the tethering becomes loose or untethered, the fundamental link still exists. The TDVP model allows for the interfacing within, across and between multiple dimensions of finite subreality (by a process called "indivension" and a content of "vortices") and allows, too, for individuals or any groups*

("individual units") to experience their own unique reality. This finite reality bridges an all-pervasive infinite subreality essence of all-embracing time, space, and extended consciousness (meaningful information) as well as extropy (multidimensional order) and potential life.

Comparing TDVP with 20 other Theories of Everything across 35 different parameters, TDVP scores 35/35. No other model scores above 20/35 other than the original models of Neppe (Vortex N-dimensionalism and Close (Transcendental Physics). But TDVP is far more than a theoretical model. It is supported empirically, has areas of testability in our 3S-1t (three spatial dimensions with one point of time) domain, has mathematical and logical support including the calculus of distinctions and generates six hundred ideas, speculations, hypotheses and extensions for research. Philosophically, the model of TDVP uniquely involves "Unified Monism".

Something inside me completely resonated with the concept, even though the scientific jargon throughout the paper was a little challenging to comprehend.

Contextually it makes sense that everything is connected in some way. It's the 'how' we are bereft of understanding so far. I have the faith, love and trust that it will be revealed as we learn to coalesce rather than compete in our chosen lifestyles on Earth.

There are some concerns regarding misuse or even misunderstanding certain things, like the Schumann Resonance for instance. Many are confused and attribute it to a frequency shift in the Earth itself.

To clarify, from Wikipedia: The **Schumann resonances** (SR) are a set of spectrum peaks in the extremely low frequency (ELF) portion of the Earth's electromagnetic field spectrum. **Schumann resonances** are global electromagnetic **resonances**, generated and excited by lightning discharges in the cavity formed by the Earth's surface and the ionosphere.

This Schumann Resonance had nothing to do with the Earth itself yet there are those who have convinced the lesser-informed that it does. It isn't the Earth; it's the cavity above it between the surface and the ionosphere, in other words.

Still, we are on the verge of a experiencing a Cosmic Evolution in the opinion of many. If that were so, many of the various schools of thought would start to say the same thing, or at least those who cross-reference resources would begin to see patterns emerging. I continued to have lucid dreams and visions that made it nearly impossible to distinguish the difference between them and reality as I knew it. It tends to make one question their sanity.

I'd had this happen in college and didn't feel prepared to handle it then, so I wasn't sure I could now. Nevertheless, it was happening.

Answering Backward Questions

I have spent a lifetime questioning the nature of reality, mostly due to the rather bizarre 'experience set' I've had. Until recently, I didn't really have someone close to investigate the questions, answers I've perceived and the continuing cosmic conundrums I encounter. If we are indeed one and there are potential perfect partners, how do we find each other?

In a serendipitous moment, a sublime flow of recognition led to what can only be described as a divine conspiracy. This conspiracy brought two people together from opposite sides of the proverbial fence, Russia and America, in a profound statement toward harmony among people and planet.

Our relationship seemed timeless and within a year we were married on the base of Bell Rock, just outside Sedona, AZ. Our discussions range from classical music's influence on society (Luba is a conservatory-trained pianist as well as performer and teacher), to the nature of being a zeitgeist for a new living awareness evolving naturally in many.

We even had a moment on our wedding eve where her Arcturian (tall almond- eyed grey) 'friend' came to visit, sharing with her that they were very pleased with our union. Luba still hasn't come to complete acceptance of her connection, yet things do happen and she feels safe enough to share those events with me. It will be interesting to see what the future holds.

Back to the spheres I described earlier, supreme in consciousness as it seemed, and the foundation of an exploration into theory. Apparently they were responsible for the creation of humans and our solar system. No doubt they explained it simply rather than give me the full story... if they did one; then it's likely they did many.

Still, the idea that three immensely conscious spheres (or at least a plausible representation I could handle) managing creation is both corroborating the age- old 'trinity' presence in Earth's religions as well as indicating that their essence could be contained in the micro-version... perhaps the proton, electron and neutron? The 666 would resolve to the carbon atom.

Our models are getting finer and finer with the discovery and/or descriptions within particle physics. However, they are the result of the deconstruction of matter, an obvious dichotomy to the creation of such. Just because we can take something apart by force doesn't mean we can or will understand how it is put together from a scientific and/or spiritual point of view.

The idea of consciousness residing in the space between the gross particles (proton, electron, neutron), and managing the formation of elements through some kind of cosmic alchemy (intelligence) is certainly worth consideration. The further validation in the TDVP, or at least its apparent tethering consciousness, space and time may also indicate this operational consciousness present in every atom and molecule. But where does that put us?

I have a tendency to pull from many resources, scientific and spiritual, to exercise my choice in accepting what resonates deeply. I think too often we limit our views to what we're comfortable or familiar. I don't often experience the deep stillness necessary to inquire and listen, yet I probably do it more than most as it happens quickly when it does.

I carry constant questions that, in my spare moments, I continue to contemplate. One of them, early on, was where the word 'satan' originated. I wasn't buying the whole 'devil' thing at all. It didn't resonate. What I found in an expanded university double-volume dictionary, you know the one that each volume is close to four inches thick, in the very first reference – Greek 'thetan' meaning 'thinker.' Where else have we been confused or mislead?

I picked up a Chick Publication one day when I was working as a machinist for Garrett AirResearch, it was on the counter at the tool crib where we got our fixtures and tooling for jobs. On the cover was something about the 66 books in the Bible, which I automatically thought about 'interpreted through the 6th sense. Numbers can be contextually manipulated as well.

I was in my mid-20s and had a very active loose association capacity, still do for that matter. It took me deeper though, thinking about the number 666 and how it has been misunderstood, but I didn't know just how. Then it occurred to me that the carbon atom was the key, having 6 protons, 6 electrons and 6 neutrons. It is the number of man, rather humanity. We're all carbon-based beings on this planet. All the separative notions don't begin from the basic understanding that scientifically, we're all ONE.

Does this understanding eliminate our separative notions and bring peace to the planet? Certainly not, but it does give us a foundation to start consider things from a different perspective. Perhaps the question, "If we are ONE, then how is it reflected in the natural order of creation into which we are intricately woven?" Where's that spooky action at a distance?

This would require extreme observation without attempts to fit the results into any kind of framework or mold, simply because we haven't had one that would contain this new living awareness in its fullness. To date it has been a subjective view, yet the sensations that go along with it often defy our ability to articulate them in any kind of objective fashion. Still, there is this thread of common experience, insight and transcendence in the thoughtmosphere that is resonant frequency we all feel.

The greatest achievement in Ufology today is that of recognizing there is a very high-level, even high-frequency, activity that is engaging and inspiring a new living awareness through the interactions of human and non-human intelligence in its myriad forms. The construct we have been missing is so perfectly displayed in front of us that we have been taking it for granted, forgetting how interactive and precious it is.

We haven't considered the nature of our world and the intricately woven thread of consciousness woven through it by beings we can't even begin to comprehend yet, this ability to listen or perceive is flooded by responses all around us that have been too subtle to perceive in our gross minds. Contact, across the modalities, makes us more sensitive. It changes our individual frequencies sublimely and subtly, allowing us to feel more of our BEing. That doesn't mean we know what to do with those feelings and sensations just yet.

Let's go back to the 'message' I got from the spheres for a moment, "We are not only your forefathers; we are the forefathers of your solar system." Let that sink in for a moment. It is replete with 'oneness' of creation, right? So how does that 'One' speak through or show up in so many?

That's the beginning of understanding the science the flows through the form, fit and function of all things... including us. We're just a bit out of touch with the science, still. A recent movie, Time of the Sixth Sun, brings so many threads of understanding and creates a unified field that is impenetrable to lower vibratory activity.

What does that mean?

In looking at the full gamut of the electromagnetic spectrum, we find ourselves at the low end of the frequency range. Our multidimensional neighbors and non-human intelligences are sensitive to these frequencies; even reside in realms far higher than our own so it seems. We sense the vibrational shift immediately before contact happens.

It's their consciousness that allowed them to evolve to that place or dimension. The natural path is to move upward on the scale, right? It's completely different for everyone and the same for everyone; a paradox of individual evolution and acquiring the permissions to access the consciousness. We simply cannot take it by storm and expect a welcome.

We all know, or at least suspect, that the hundredth monkey paradigm will cause massive changes in humans as we continue to move toward the balance of the spiritual and material worlds, the non-human and human beings in a sublime dance. The music takes us toward seeing the world through new eyes, new thinking and new understanding as the dance progresses.

Although it's a real challenge for most of us to think of a true new world order of that magnitude, where the natural abilities, awareness, skills and talents bubble up to the surface as we seek them within. Like Einstein has been quoted as saying, "You cannot fix problems with the same thinking that created them."

Is a unified field even possible to such a degree in our lifetimes? Perhaps It will be. It's been said that when choices are made, the world changes before us as if the universe is driven to fulfill the intentions of a pure heart. Now there are those who would argue that the Law of One places us in two choices, service to self and service to others.

The obvious notion being the greatest service to self is service to others, like being selfishly selfless. Getting out of our own way is the key to allowing this new awareness to emerge from within us. Every known modality points to the same notion – fearless living moves us toward learning how to achieve harmony among people and planet as a holistic system with all the kingdoms involved.

Perhaps the crop circle on the next page and its decoded message offers food for thought. Who are the deceivers that are being opposed? What conduit is closing? A conduit indicates a structure that has something flowing through it, right? Have we been anesthetized to Mother Earth and Nature's flow for too long? See how we can skew thinking in context with simple questions?

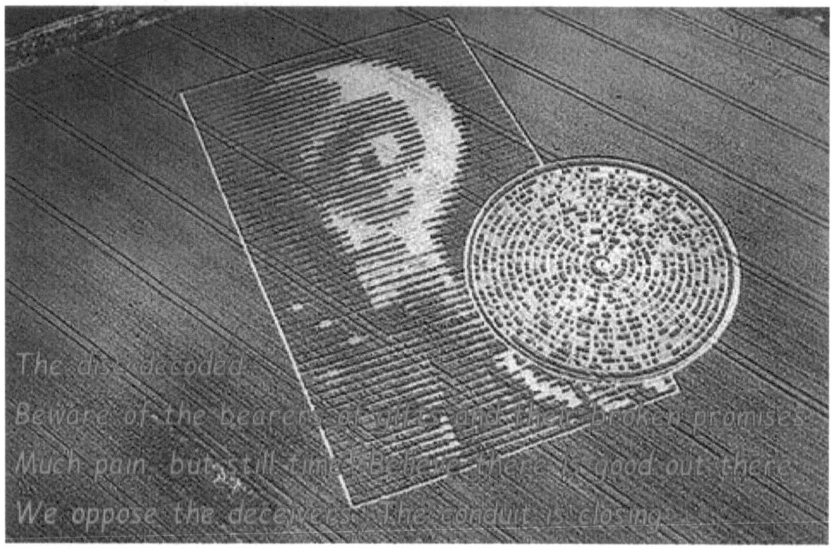

Congruence of the Contact Modalities

A theory is only as good as the elements one can find in reality that support it. For the contact modalities to provide a consistent and/or interrelated presence in our lives we need examples to compare them. This is a challenge because we all interpret our perceptions slightly differently with a different voice uniquely our own. However, there ought to be something that gives us a common experience across those individualized perceptions.

There are possibilities across the various contact modalities that we can explore here. For example, they include Channeling, Lucid Dreams, Mystical Meditation, Near Death Experiences (NDEs), Out of Body Experiences (OBEs), Remote Viewing, Spiritually Transformative Experience (STE), Shamanic Hallucinogenic Experiences and UFO visitations. In order to have a foundation for comparison, let us look at some identifiable traits among them.

Channeling – In some circles there is a consistent experience of information shared regarding the transformation of the consciousness and therefore the civilization of planet Earth. It has been my experience that many channelers present information from a higher aspect of their own consciousness and still continue to engage ideas about the dissolution of 'cabals' and the like. Good intentions and old belief systems seem to filter through in many, but not all, cases.

If one is able to be still (free of thinking) and feel the information (without prejudice) as it is presented, there is a noticeable difference between the good- intentioned and the actual connection with other beings. This is one of the changes in awareness that happens as we progress. It has been this one's experience that message of importance speak directly to the evolution of the specie (humans) and not toward events or other agencies operating on Earth (cabals, et al). The ability to observe without projection of belief systems is most challenging.

Lucid Dreams – Individual consciousness is affected, even engaged, by this type of experience. Lucid dreams don't always happen at night, they can happen in a flash during any time of day. Some might experience a moment of lucid interaction when they close their eyes and have a vivid display or even an engaging interaction with one or a group that has a significant moment of inspired activity, something that has personal significance toward activity in the world or it might provide an answer to a question that has been stirring around in one's mind, for instance.

Mystical Meditation – These are experiences that usually involve a prolonged chant or focus with music that provides a gateway into inner realms normally unavailable to the waking mind. Common experiences include speaking with Masters or observing displays of terrestrial or even cosmic scenes that initiate an inspired sense of well-being or even transcendence for a short time. Most often these experiences lead to a sense of oneness beyond comprehension.

Near-Death Experiences (NDE) – These are usually triggered by a catastrophic event that can include a traumatic event (car crash), during surgery or even an emotionally traumatic event. Most seem to include a profound sense of connection with either relatives or spiritual beings. The return often brings a sense of awe and a jumbled onslaught of profound sensations and thoughts toward a sense of profound fearlessness and freedom. It often leads to a realization of an expanded consciousness that had been previously unexplored and even unrecognized.

Out of Body Experiences (OBE) – These can begin in childhood and extend throughout one's life. Some experienced them regularly and some intermittently, spontaneously in the moments just before and just after sleep. Occasionally they are also experienced by those who engage some form of ethnogens, a substance ingested to produce a heightened state of awareness and/or an exploration into inner space. This experience also provides a sense of profound connection to a consciousness beyond human, or at least perceived as being beyond human.

Remote Viewing – This is a practiced form of the OBEs that has been around for some time, although more recent activity by government and military concerns might give the impression that it is a 'new' technology. Ancient texts have often referred to visions of certain people and places as prophecies or revelations. It may not be identical, yet the similarities are striking. The resulting sensation is also one of profound connection to awareness beyond human.

Spiritually Transformative Experiences (STE) – These events span the gamut of significant emotional events to journeys in the wilderness that provide seeming paranormal activities that leave the experiencer with a profound sense of connection to people and nature, often beyond the perceived human capacity. Many life-transforming events produce a spiritually transformed outlook on personal understanding, life and the world.

Shamanic Hallucinogenic Experiences – These can happen with or without the presence of a shaman or spiritual guide, yet the important thing to note is that most are completely inner/internal experiences with eyes closed. There are rare events where there is no difference with the sight, whether eyes open and eyes closed the same thing is 'seen.' These also produce a profound awareness of connection to consciousness beyond the normal human scope.

UFO Visitations – These are most varied, from actual contact (physical) to multidimensional (conscious and physical activity beyond normal physical constraints, like moving through walls et al) to interactions on craft or

telepathic communication to 'downloads' of information received in rapid fire succession. Often the events begin with at least a slight sense of concern or fear and transform over time to a profound sense of cosmic affiliation beyond human.

There seems to be a consistency in the elevated sense of one's ability to perceive that is strewn across all the above. In most cases, it seems that one does not have just one type of experience throughout their lives. We each engage, or are engaged by, various kinds of circumstances and situations where multiple methods may be included in one primary experience.

The stigma in the general public is that these things are beyond logic and/or reason and therefore suspicious in their ability to evoke or impart knowledge and wisdom. Still, from the subjective responses reported, these experiences all provide insight and inspiration for advancing self-awareness. It is potentially leading us to self- actualization and self-realization.

Before closing, let us look at self-actualization and self-realization as potential results of extended engagement of the contact modalities. If we have a personal destiny of some kind, it would probably include our 'perfected' form, fit and function in the world.

Would you agree, at least in theory, we have a 'purpose'?

In that sense, self-actualization is the epitome of applying one's skills set to reach their full potential personally and professionally. So also is the nature of self-realization the epitome of one's inner development and reunion with the unified consciousness from which we all descend into this playing field we call Earth in these bodies we call human.

Wise men and women continue
to grow through altering their
awareness and belief systems as
discoveries are made, especially
those that raise new questions
about habitat, health, safety and
sustainability. We learn to work
together by following the wise
movements, incorporating them
and sharing them throughout
existing systems in the world today.
It is imperative to do so.

Continuing Enigma

I've grown my knowledge base in answering these questions beyond expectation. I believe our understanding of science and spirituality is in the process of merging and sharing these words only edifies that in my heart and mind. The two worlds necessarily need to have cross-overs and even merge as we grow in our knowledge and ability to see patterns everywhere. Truth is like fractals that are present everywhere, just not always visible with the tools we've used in our paths to discovery of how things work in harmony. Perhaps we haven't even looked at how things work in harmony as our tendency is to look for differences and get caught up in them.

If there is a Theory of Everything that bridges the worlds of science and spirituality, then perhaps we'll see some reflections in society and our capacity for learning how to get along better. Regardless of the contact modalities bridging the worlds of experiencers into a unified field, we still need to have practical and pragmatic actions. We've been asked to step up our game in nearly every way imaginable from all the sources of non-human intelligence as far as I can tell; addressing our sustainability.

Considering the population of Earth and the experiencers of contact modalities, one or many, the nature of them seems to lead us toward a sense of community, a greater unity than we are currently experiencing. Perhaps this is just a natural progression of a planetary civilization. Not all inhabitants of the planet need to experience a profound relationship with non-human or even their own consciousness. The relationship between science and spirituality, once at odds, is gradually showing signs of congruence, recognizing that definitions and explanations are describing the same aspects of personal and planetary evolution in different language.

Consider the Kardashev scale for a moment. (re: https://en.wikipedia.org/wiki/Kardashev_scale) If we look at the types of civilizations in comparison to the types of contact we experience, there is likely to be a variety that corresponds. For instance, perhaps the physical ships are from advanced Type I. Type II may include the organically integrated versions of ships some have encountered. Type III, a galactic level, might also have something to do with the Galactic Federation and the administration of worlds in embryo, like ours. It is just another way of looking at the same events and offering additional considerations that increase our capacity for intelligent engagement and further understanding.

The frequencies many hear are another potential unification component. We know the electromagnetic spectrum is vibration, yet we haven't really considered that thoughts also have frequency. We know that data packets have a frequency, too. Perhaps universal thoughts are specific in their frequency distribution and accessible to those who become adept at meditation or journeying facilitated with ethnogens. They seem to help, yet the obvious prudent path is to get there on your own.

Manna from heaven might have been the first reference to ethnogens in spiritual literature. Richard Alpert, whom we know as Ram Dass, found that LSD is a method for attaining the Eastern spiritual realms without the rigors of years of meditation. DMT is also know to produce short-term enlightened moments, though the reality is we have to learn to be present without any exterior catalyst in order to anchor our awareness in daily living. That is what it is about... daily living. The latter requires a presence of mind, unaltered by anything external.

I believe the study of and/or understanding that these subtle frequencies present in the substrate of every experience, if we are cognizant and observant; will lead us to further communication in ways we never imagined possible. I know this to be true for me, so it isn't just a theory or whimsical notion. I live in this world and yet so far beyond it. There is an experience of physics, for lack of a better, that resolves into a new living awareness. It's often preceded by a synchronicity that one recognizes, perhaps pauses for a moment of silent recognition, and opens therein opens the door to an insightful thought that bubbles up from within.

Often when we hear something for the first time and it sounds foreign. Our first notion is to reject it regardless of its truth, especially if we've never heard it yet. The degree to which our perceptions and world views are involved also affect our ability to allow the disparities to coalesce instead of create a cacophony of confused thoughts and cognitive dissonance.

We often don't consider that the communication of another is congruent with our understanding, only the words are different and we tend to feel a loss of flow, rather than pause and reflect, allowing the similarity to emerge naturally. Sometimes there are those who make those connections and bridge the separative notions in populations, whether it is within a specific community, ethnic groups or nations. We grow together as a result.

I often wonder if we're approaching it like the reverse-engineer craft we see in our skies most often now. Instead of perceiving there are connections to everything and seeking them, which tends to 'push' energy outward. Might we look from the other end of the kaleidoscope and just observe rather than look?

If, as much has been written toward, everything is ONE then how might that fact reflect in the observation of our lives and world? Asking those kinds of questions will also produce answers, but they may not be from the same thinking that created the question. The advancement in consciousness, as we are finding, often takes us into realms previously unexplored, let alone experienced.

The contact modalities, human and non-human intelligence and the consciousness that drives the bus we're all in, coalesce into a new living awareness for us to experience if and when we are ready. Is there a New World Order that will emerge? Perhaps it is more of a newly ordered world. Will we be able to self-organize into fractals of a greater holistic system, differentiated by the relevant technologies and distribution networks that we need in order to live sustainably on Earth? We might want to consider the distractions of fear-based rhetoric in the population and change its course.

Could there be a Peer-2-Peer Collective Intelligence Platform that will emerge through the efforts of those already taking action? It has been my assertion for decades that conversations that matter, of higher order, are like seeds in the thoughtmosphere and those who engage them have access to those who've gone before and/or are present today.

It is yet another level of contact modalities we don't necessarily consider. There is such a magnificent pool of information that is accessible to those who venture inward and upward, fearlessly exploring the possibilities of engaging non-human intelligence or even Divine Intelligence as we learn more about our capacity for co-creation. We already co-create. Unfortunately we've not been too conscious of our creations.

Collectively, is it possible for us to engage the systems theory that brought about the first photo of a black hole to create globally integrated systems that serve humankind? Computer systems all over the globe were networked to share data that ultimately gave us our first look at a black hole. Imagine what networked minds could do. Individually, is it possible for us to acknowledge and respond to our 'sameness' instead of our perceived differences? Time will tell.

As Jose Arguelles (RIP) said to me years ago, "The Galactic Federation looks forward to the New Time!"

Alas, the sense of community is contracted, fragmented into a thousand pieces across the fields of study as it seems folks have such tunnel-vision that they aren't looking for correlations across other fields. The soil is all the same. It is Earth after all. The crops are different, yet they all feed the world as we know it. I've staked my life on our ability to unite as one people and one planet in my lifetime. Just believing it is possible is a step, removing the distractions and doubt is another.

Observing your intentions and distractions is another, it is where faith grows and trust follows. We're all the same, we all just want to love and be loved; to feel safe, secure, supported and sustained in our growth toward harmony among people and planet.

If this chapter has helped correlate or corroborate your own thinking, even in some small way, I've been successful. If your own awareness had broadened and your thinking now includes greater capacity, then I've been tremendously successful. I know I'll learn more than I could imagine from reading the published book with all the authors chapter contributions.

This book is in your hands now, too. Let your gut be your guide on your quest for truth and understanding. Question everything. Even with what I have experienced and know, I still question my conclusions as well as the experiences themselves. Every time I do, I learn something new.

Are We Waking Up?

"Awakening is not a matter of intelligence. Many intelligent people are deep in illusion. But rather, awakening happens when one has an insatiable thirst to know the truth." *Author Unknown*

Awakening or illusion? How that insatiable thirst dispels the illusion depends on how awake we choose to be and the willingness to suspend or slice away false belief systems. We have to test them first, of course. Recycling memories tends to offer new insights and thoughts for me. Perhaps they'll be interesting to you as the stream of consciousness flows onto the page. Putting thoughts to writing can be so cathartic.

What is truth? We refer to it, but do we ever really define it so that we can agree on it? I mean, truth would be ubiquitous and undeniable, right? I suppose that might be a clear explanation of how we work, how we relate to each other, how we relate to the world or worlds we may perceive?

Perhaps a combination of a mathematical and psycho-spiritual model would explain the intelligence at work throughout? What is awakening related to this?

What would it do if we knew it? How would our lives look if we not only knew it, but lived it? Would you say that within this insatiable thirst there is a new living awareness that emerges? What is that like? Does it make sense? Is that sense being made common?

This emerging awareness seems completely subjective in the discovery process. Attempts to articulate a multi-dimensional funnel of data into linear ears and eyes has been going on for millennia. Alas, we still tend to argue about the details and miss the overarching message(s). Can the insatiable curiosity in many cause an upwising in the world? Big T, little t, it's ultimately the same, so what is your concept of truth?

Well, that got a little deeper than intended. On a more practical note, it sure seems that this theme runs across a plethora of playing fields in our present time. The information age and data available for searching is causing an evolution in intelligence across the world. It would seem only logical that the ability to discern would also reach new levels in our lives.

Awakening - Illusion

Once we find this 'whatever we're looking for' truth, the mere acquisition of it might make it suspicious at first. It's natural. We're always confused by the unknown suddenly becoming known, which it would seem is how the process of discovering truth works. Applying that to the field of Ufology is no small task. It's hard to cut through the din.

There is a lot of concern voiced by many regarding the 'experts' in the field and the way in which they are promoted, or not as the case may be. There are numerous questions flying around social media platforms in the hundreds of groups that are aligned under the Ufology banner. They range from the absurd to the transformational in scope, but there is little interaction.

It's fairly easy to spot the absurd or pejorative and dismiss them as just uninformed or experientially bereft. Then there are those who have great stories and no real proof, yet are followed by thousands or possibly millions across the web. Let us not forget the researchers who have done a great job at collecting and sharing information, though they tend to speculate about potentials based on incomplete data; experiencers are left out of the loop.

It's been less than a decade since experiencers, en masse, have started sharing their stories through blogs, groups, pages, podcasts and web chats. Less than 20 years ago when the curator first dove into the web, there were extremely small numbers of folks willing to expose themselves to scrutiny. Spirit Web was big then, a chat platform with multiple rooms and topics.

Today, the scene is prolific with thousands of opportunities to engage on hundreds of platforms.

The Aerial Phenomena Research Organization started in Sturgeon Bay, WI in 1952 and moved to Tucson, AZ in 1960. The National UFO Conference began in 1963 in Hollywood, of course. MUFON, Inc. was originally established as the Midwest UFO Network on 31 May 1969, in Quincy, Illinois. The conference craze began emerging in the early 1990s with the first International UFO Congress in Tucson, then Tim Beckley's *New Age and Alien Agenda* conferences.

The Prophets Conferences were big in the mid to late-90s, the curator managed their Phoenix event in 1997. Whole Life Expos began around the same time, too. An ex-board member of WLE started Contact in the Desert, another profitable venture, in 2013.

Awakening the Fray

Now there are dozens of Ufology-focused events around the world, most of which can be found on our website as we become aware of them. It is a niche market for sure, yet it has its celebrities and movie stars, just like the old days. Just how those celebrities are made is another perplexing question as there are many agendas spiraling within the field; some glaring and some a little less detectable at first. There are those who genuinely admit they don't know much, though they have managed to gather enough data and some objective views on the subjective experiences of others.

We tend to listen and nod our heads to the experts of pretty much any field because we rarely understand what they are actually saying, especially the Ph.D.s, yet we don't want to appear stupid to our peers by asking questions. Those who do have the experience and/or intelligence to ask questions are generally felt to be critical instead of sincere. Intelligent questions get glossed over and go unanswered yet tend to throw up those BS meter flags, or at least they should due to the avoidance.

We certainly do not have all the answers and maybe not even the questions yet. We still need that insatiable thirst for truth. We've had a materialist quest for truth with evidence, exotic metals, photos, ship remnants; material proof of some kind. We have only asked cursory questions regarding the occupants of the crafts we either have seen or trust that are there. The questions cannot be asked because the experiencers have not been let into the game.

Like any process and reflective of the field in general, there are some pretty strange tales and loud mouths in the experiencer groups, too. There are some consistencies, though, however small, that might be a good place to start asking questions. The quest for truth, at any level, is humbling beyond imagination. The small voices that seem genuine and sincere may hold some deeper truths about our relationship with the other planetary civilizations that frequent here.

Humans are resilient, resourceful and even reverent toward challenges in chaos and turmoil. We show as a human race that we can unify to overcome them and restore order. We do it all the time, yet somehow get distracted when it comes to the onslaught of information, personalities and pundits who manage the gate. There are some new developments within the field that are leaning toward incorporating experiencer data and philosophies on contact.

The key for the expansion and potential unification of this field is in collaboration with intention to keep the attention on asking better questions. Like Einstein claims, we cannot answer the questions with our own thinking, yet it is our option to be open to new thinking that allows the flow of information. Many have events where they speak things they did not understand prior to speaking about them in that moment. There is a conduit for information flow within us.

What does that have to do with the insatiable thirst for truth? That flow is what quenches the thirst. It's like the elixir that resonates throughout our body/mind/spirit/soul to the degree of which we feel 'electric' or 'energized' or even 'on fire' with this discovery. That is the awakening of the conduit within us.

This awakening brings humility, not arrogance; compassion, not surreptitiousness. As curator and partial purveyor of profundities, we can all work together to restore order, the most profound truth of all.

Highlight of Recent Events

In recent years I was led, through a series of serendipitous synchronicities to the love of my life. That experience alone was profound. Luba gifted me with a subscription to Ancestry.com which led to the discovery of my birth parents. For me, this is a direct reflection of the years of trials and tribulations that were worth experiencing in order to find the truth of my life.

I was invited to attend an outdoor music festival by a friend who was the front man for a band called Return to Zero. Their music was a bit intense for indoor listening, so this gave me the opportunity to support them in a more tolerable environment. Arriving at the event I soon encountered a young boy, perhaps 5, that immediately started talking to me about his experience on the playground adjacent to some park benches, where I took a seat to listen to him.

A few moments later his father walked up while at the same time my friend approached from the other direction. My friend immediately called out to the boy's father and told him I was the guy he wanted to introduce. The three of us talked for a few minutes.

I sat down with Brent, the father, and talked with him for nearly two hours. In the conversation he told me about his recent upheaval in life and a Kundalini Yoga teacher training program he was involved in that was just kicking his butt.

A week after the event I got a text inviting me to his graduation. It was some distance away and I normally wouldn't have gone, but something inside felt like it was important. I agreed to attend. I didn't think anything more at the time.

A few weeks prior to the music event I had made the decision that I was just fine being alone, giving up on the notion that I would find a perfect partner that would share my passion for the unknown and understand my eccentricities. A few days prior to the event I had a dream where I hugged a woman, even though I could not distinguish her features, and felt like we merged as one being. The sensation was so intense I woke up immediately, and then returned to the dream state to find that feeling again.

I returned to that sensation multiple times over the next few days, wondering if it was a sign that someone was going to appear in my life and hopefully soon. On the way to the Kundalini training graduation I considered the environment being a perfect place to meet someone that would at least be compatible.

Upon arriving I heard a familiar voice call out my name. It was a member of a discussion group on UFOs that I had facilitated for a short time the previous year. We hugged and talked for a while. She had just finished a class, told me she felt there was someone significant I was going to meet and bid me adieu.

I went over to the group collecting outside the building, enjoyed some ginger tea, and waited for Brent and his son to arrive. They showed up just as we were being asked to come inside for the ceremony. I went in with Brent and his son and took up a seat in the back of the room. It was October 28th, 2016.

In a few moments a woman crossed in front of me, quite attractive and looking about my oldest daughter's age which made me feel a little uncomfortable. I felt uncomfortable because as she walked in front of me, with no eye contact, my heart literally felt like it flipped or spun in my chest for a moment. It was something I'd never felt before and, of course, put me on notice to pay attention to what was taking place.

Long story short, I introduced myself to this amazing Russian woman after the ceremony was over. It was like I'd known her forever and, after my introduction all I could do was gaze into her eyes. Her first words, "Are you reading me?" brought me back to reality and I apologized for my non-blinking gaze. She actually liked it and wasn't uncomfortable in the slightest, mentioning that she felt something going on inside her as well.

I moved her, her 12-year old son and 83-year old mother (who spoke only Russian) into my home just over a month later. We were married on the base of Bell Rock, just outside Sedona, on the fall equinox of 2017. That Christmas, Luba gifted me with the Ancestry subscription, hoping that I would find my birth parents. My heritage, based on my experiences and the impressions of some others, was questionable still. I needed proof.

In January of 2019, I got my proof. My half-sister had registered and a notice was sent to my email about a first cousin or half-sibling match. Within a week I'd connected with my birth mother as well, enjoying hours of conversations and video chats over the next few weeks and months. We traveled to Indiana and stayed with her, meeting my half-siblings, some of my other relatives and celebrating my 62nd birthday with her.

It was obvious from the beginning that the connection was real. Even our personalities and styles of communication were strikingly similar. Luba took a picture of us in as we looked into each other's eyes and our profiles were nearly identical; further evidence. Our week with her was as wonderful as it could have ever been, and her regrets and emotional turmoil subsided even more.

She'd made the best decision she could at the time, having a 4 year old already and no husband to rely on. The two had met while she was working at his diner, My Place, in Anderson, Indiana. The relationship flourished for a while until she got pregnant and my maternal grandmother put the fear of God in him.

She told him if he wasn't going to marry her he need to leave the area pronto. I suppose he considered his options and just wasn't ready. Apparently he was divorced once already. Dad had left just a few weeks prior to me being born, citing he was going to pursue an acting career. She was still a bit angry, but those feelings subsided the more we got to know each other.

Now here's where it gets spooky cool. Just after my divorce I had prayed to find my birth parent yet again, only a little stronger because of my depressed state at the time. A little later, I'm not sure if it was weeks or months; I met an older guy at a UFO discussion group and became fast friends. His name was Lee Carlile. He lived in Groom Creek, Arizona, just outside Prescott and invited me to spend the weekend at his cabin sometime.

I knew some folks in Prescott who ran groups and a week or so later Lee called me and invited me up to one of them. He had a house with a cabin next to it, living in the cabin and renting out the house at the time. I don't recall if we ever talked about where we were from originally, so the question of paternity never came up at all.

After reconnecting with Mom, though, I found out my father's name and began trying to track him down. It was Leroy Carlile. The thought of it being the same guy I'd met 30 years prior was both exciting and felt nearly incredulous, especially since he was headed to New York for acting, or so Mom thought.

As it turned out, it was him. The events and research unfolded with the help of a childhood friend who had the search software and wanted to help. Lee had passed a couple of years ago to my chagrin. We'd made several trips to Prescott to try to find him, even having a last known address at an assisted living community.

His Facebook profile eventually showed up, though I didn't find it upon initial searches. Although it didn't look like it had been active for a while I did find some folks that appeared to know him well and reached out to them.

I found the guy who was renting the house at the time and he remembered me. Imagine the odds. He told me that Lee had passed and that he was in transit from one location to another when he fell, was hospitalized and passed from complications before his records could be updated. Lee's daughter had informed him of the tragic scenario a few weeks after his transition.

My point in sharing this is to illustrate how intentions and actions produce results beyond our imagination. I've felt for some time that inner and outer realities can merge in practical ways. My life seems to exemplify the reality.

I go so far as to say that what mystics have known for millennia – intention, attention and action – produce or manifest a new reality and reflect the experience of what quantum theorists have posited as the process of observation converting waves to particles. We can create our reality.

I'll go further to say that the essence of the communications we receive from non-human intelligence also edifies that we create our reality. We are so unaware that we do, that our feelings and thoughts have produced a world in chaos.

We can, in my humble opinion, change the course with awareness, communication and collaboration with conscience.

"The best teachers are those
who show you where to look,
but don't tell you what to see."

Alexandra K. Trenfor

What is Reality?

Questions about reality abound, especially regarding consciousness. What known phenomena in nature, considered emergent or otherwise, are similar to consciousness? Perhaps this is a question regarding contact experiences as well since it appears to be an emerging phenomenon.

Consciousness, like the elements of nature, is always adapting and evolving. We didn't even know it existed until we began exploring the spiritual or metaphysical nature of our existence. Humans tend to see themselves as separate from nature, even though there is evidence across the spectrum of our own evolution that we are unable or unwilling to consider.

A perfect example is the 'missing link' of human evolution. We will absolutely not consider the ability and nature of mutation. We needed a larger cranium to house the growing intelligence. The results, over time, were the mutations that brought about homo sapiens. It happened gradually over a period of time. It is the same pattern in organic life, mutations based on the need of the species to adapt to new environments, both biological and ecological. More than likely it has little to do with genetic manipulations from outside the human genome.

Educating Ourselves - Drawing Out

Those who are stuck in old paradigms, usually from the education based on previous discoveries, however misunderstood or misinterpreted they may be, tend to only see things through those lenses. Just one example of the shift in thinking was offered by Gregory Bateson in Steps to an Ecology of Mind, where he states, "What we have to think about is how the pieces of knowledge are woven together. How they help each other."

"Matter in quantum mechanics is not an inert substance but an active agent, constantly making choices between alternative possibilities.   It appears that mind, as manifested by the capacity to make choices, is to some extent inherent in every electron. This does not mean that an atom has the same consciousness as a human being, but rather that an atom has a reflective capacity appropriate to its form and function."
physicist Freeman Dyson

Vernon Neppe and Edward Close presented yet another perspective in their TDVP model, where consciousness, space and time are tethered across nine dimensions. "The discrete metafinite (finite and transfinite) is embedded in the continuous infinite. Discrete refers to the measurable pieces. They are like quanta, discrete, pixilated elements. But each element can be measured in Space, Time and Consciousness. All aspects of Reality obey the Laws of Nature."

Further Considerations

After some considerations, I've learned that an experience system within the contact scenario, the collection of sensory perceptions gathered over an array of intentional and occasionally unintentional events, develops through *feeling things out*. Everything is vibration, right? That includes everything in nature, which means it is, by default, a part of consciousness. It's not separate.

So we feel first, obviously, our bodies are the receptors of the signals (actually transceivers) and illustrate that nature, according to the indigenous ways, the gut feeling is our first connection with our environment or outer reality. We send from there, too. It's downright reciprocal, but we usually aren't conscious enough to realize it, let alone observe and learn.

I agree also with our innate ability, often unrecognized, of naturally reaching out to perceive 'what is' in a kind of constant reciprocal event. Structurally I think it is a synergy of chakras, meridians and 'clair-alls' (send/receive circuits) that is at work always. Again, we aren't aware enough to deepen and/or expand our moments of coherence it seems. Sometimes we are.

We often hear of the notion that we create our reality. It has occurred to me, and I wonder if, the way in which we create reality is the interaction between the positive thought (or negative), neutral object and the electromagnetism attracting loose and/or shared electrons of people, places and things that resonate. It's the spook and spooky action at a distance thing applied locally.

It seems that this is a recurring theme, yet I don't find it articulated as a consideration in discussions regarding contact and its significance. I mentioned this earlier – Attention, intention and action move matter in ways we do not understand, yet we experience as manifestations of things we've asked for, planned for, even prayed for in our lives.

Your thoughts?

It's not just what you think, how you think matters. The application of critical thinking and questioning authority is crucial for us individually and collectively. Holding on to false beliefs, separative notions and a competitive nature created by folks who, perhaps with the best of intentions, attempt to control our path is a disservice to each of us personally and our planetary society as a whole today. We simply must hit the pause button, reflect intentionally and move forward with a more compatible mindflow.

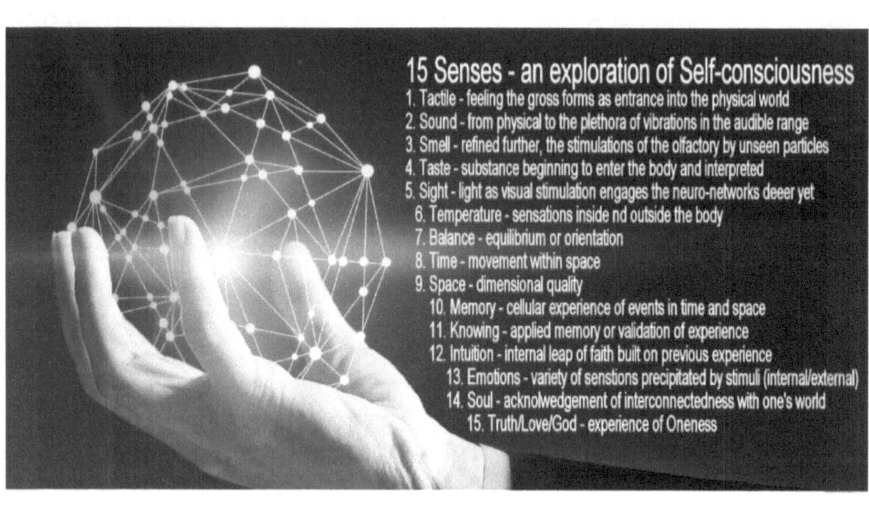

15 Senses - an exploration of Self-consciousness
1. Tactile - feeling the gross forms as entrance into the physical world
2. Sound - from physical to the plethora of vibrations in the audible range
3. Smell - refined further, the stimulations of the olfactory by unseen particles
4. Taste - substance beginning to enter the body and interpreted
5. Sight - light as visual stimulation engages the neuro-networks deeer yet
6. Temperature - sensations inside nd outside the body
7. Balance - equilibrium or orientation
8. Time - movement within space
9. Space - dimensional quality
10. Memory - cellular experience of events in time and space
11. Knowing - applied memory or validation of experience
12. Intuition - internal leap of faith built on previous experience
13. Emotions - variety of senstions precipitated by stimuli (internal/external)
14. Soul - acknolwedgement of interconnectedness with one's world
15. Truth/Love/God - experience of Oneness

Recent Discoveries and Perspectives

In March of 2021 I was introduced to the memoirs of Wilbert B. Smith, originally published in 1964, two years after his death. Wilbert was a well-respected scientist and project manager, given the task of running the Canadian Ministry of Transportation's UFO investigation program during the 1950s.

His memoirs contained chronicles of conversations he had with 'people from elsewhere' and the information they shared with him about consciousness, human beings and the reality we create. The 'book' is called *The New Science* and you could 'Google' it, I'm sure.

I'll share a couple of excerpts that made my head spin as the dots got connected in some new neural network that came with thoughts and visuals simultaneous in a part of my mind that apparently has been dormant or at least latent with potential.

Before I do, I want to note that my wife and I had already begun to ask deeper questions of how to engage this 'new living awareness' more fully to invite the experience of harmony into our lives at more intimate levels. We had a basic understanding of the possibility and some independent experiences that gave us indications of its reality, yet we hadn't been able to have a sustained relationship with it.

This is where the concept of 'twin flames' seemed to make sense on an experiential level because of the interconnectedness and reflection of multiple aspects of being as we proceeded with 'analyzing this' and 'deconstructing self' (Harry wasn't present). There were hints of inner and outer realities merging as a shared experience, which gave power to the faith, love and trust we held.

Fields of Play

Excerpt 1, The New Science, page 12

Just as Area has Length incorporated in it, and Volume has Area incorporated in it, so has the Electric Field the Tempic Field incorporated in it. Each of these three fields are mutually at right angles to each other. The three fields together are the manifestation of Reality in the Field Fabric as perceived by Awareness. The interrelationships between these various fields manifests to our Awareness as Matter and Energy, and the great variety of these manifestations is well known to us.

It often seems like the entrance to the inner planes is full of twists and turns, spins of the point of awareness that takes in the experience as a whole at first, just as a meta-cognitive multi-sensory mind/presence would do. We have no language, as of yet, that allows us the freedom to move beyond bi-polarized trigger words for describing the experience through faulty lenses of fear.

As I read the above excerpt, the following image came to mind:

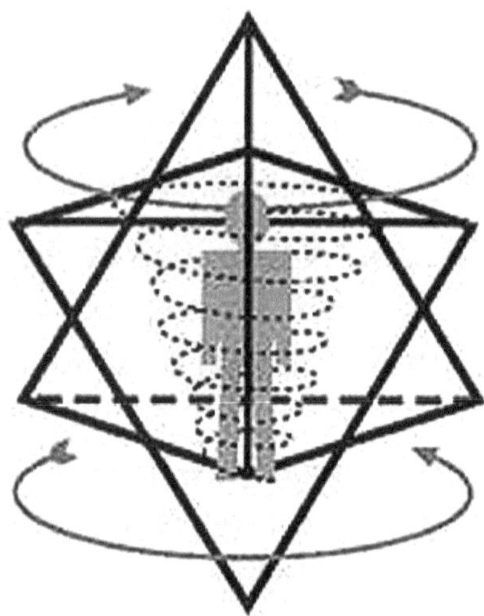

I'd long considered this model similar to the caduceus, only represented differently, with the masculine and feminine aspects revolving around the One Mind; perfected form, fit and function of the individuated form (cosmic consciousness condensed) in the world of form. All things being in order, it would only make sense.

With the caduceus, Hermes' staff, the masculine and feminine are presented as serpents entwined around the One Mind with the wings signifying ascension when all aspects are in harmony.

It might seem like round pegs in square holes, though understanding ancient teachings with new insights are ways to evolve a civilization. Hermes/Thoth presented wisdom we call the Emerald Tablets that encapsulate basic rules of engagement with reality that work most effectively, though few master them. His staff held the secrets:

At best, these two-dimensional models allude to a third. This fractal of the trinity yet again presents a potential for deeper understanding of the harmony innate to our balanced fields. We'd probably consider this important to building relationships, especially in recrafting our collective reality as we come out of Covid.

Inner and Outer Realities

Excerpt 1, page 15

It is at once apparent that there is no prescription of any "absolute" quantity of Reality, or any scale of dimensions at all in this derivation. Therefore it follows that any and all Reality will manifest Form such that its apparent boundary will manifest half of the Reality inside and half outside. We do not know, and may never know, the total amount of Reality in the Universe, but the Form of the Universe will be such that half the Reality is inside and half outside. At the other extreme, the smallest bit of Reality that we can discern will also have this characteristic of one half being inside and the other half outside.

We've had the notion of inner and outer worlds presented to us in myriad ways for millennia. This gives a new view of an old paradigm. If/when we are able to perceive such in ways that make sense and are replicable; developing the inner life becomes strikingly important and relevant in practical and pragmatic ways. The nature of how we live will change, perhaps a natural progression toward harmony with self, others and Earth. Let's consider another view.

Einstein was obsessed with figuring out how the unified field works. He had a basic understanding the quantum entanglement existed, even spooky action at a distance. He was trying to observe it, report on it mathematically instead of experiencing it. No doubt he had moments that imbued his theory with a bit of experience.

Recent reports and understanding of how the Law of Attraction works, just one label among many, is applicable for our ability to invoke the waves of intention that eventually manifest as the particles we observe in people, places and things aligned. I suspect this hints at a deeper discovery available by experience only.

I posit that quantum entanglement with the unified field as a personal experience can be at once identified by 'synchronicity' in that multiple streams, themes and waves coalesce into the 'moment.' We often don't give it much attention beyond the moment, as it leaves us with sense of awe and usually joy. The enjoyment is all we need.

It's great to observe and report as a personal subjective experience. It is magnified when two or more can observe and report the same or similar experience of that moment. That happens quite often, actually, and goes under reported because it is so common. In truth, or perhaps, it's just a 'moment' of bleed-through from the established inner connections already in existence. Is there more?

During this period of history, with obsession on self-hygiene and sequestration that began with Covid, perhaps this outer view can turn inward and a 'cleaning up' of our emotional, intellectual and spiritual worlds will ensue. There seems to be a reflection in at least some that this process is indeed happening; conversations that matter deeply are more present and in public view via social media. The virtual world is opening global conversations and interaction.

Spherical Thinking

Excerpt 3, The New Science, page 16

The following analogy may assist in visualizing to some extent the relative structure of the universal parameters and Awareness. Consider a sphere which is completely surrounded by twelve other spheres of the same size. All the spheres are in contact, and all the outer spheres are in contact with the inner. Only three spheres and the inner sphere can have mutual contact under any one circumstance.

Radius of these three spheres may be mutually at right angles and the radius of the inner sphere may be in line with any one of these. Depending on the starting point and progression, all spheres may be inspected sequentially, without any preference. A quadrature relationship is possible between any three spheres, including the central one. All these relationships are valid with respect to the relationships existing among the twelve parameters and Awareness.

In context with the rest of the material, there is an unpacking of detail we've observed in much of quantum theory already. It even aligns with the Triadic Dimensional Distinction Vortical Paradigm of Neppe and Close.

'They' [people from elsewhere explain the twelve parameters, though I recommend the reading of the material thoroughly to get the full story, in this configuration:

1. Length

2. Area

3. Density

4. Change

5. Divergence

6. Curl

7. Random

8. Decision

9. Sequence

10. Form

11. Multiplicity

12. Aggregation

Much of the expression of inner planes has been in the form

of drawings and pictures of what some refer to as 'sacred geometry' and yet the following 'Metatron's Cube' looks eerily familiar given our ability to unpack that last excerpt. It looks like this; though use your imagination for a 3-d perspective:

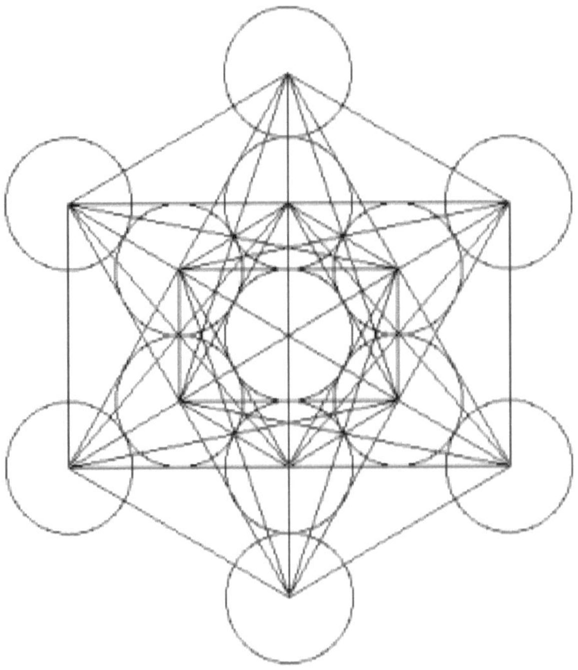

The Fifteen Fundamental Properties of Wholeness

The degree of life which appears in a thing depends upon the life of its component centers and their density. Following are fifteen structural features which Christopher Alexander has identified as appearing again and again in things which have life.

Together, these fifteen properties identify the character of living systems. They are the principal ways in which centers can be strengthened by other centers. They are not independent, but rather rely on and reinforce each other. Things which are more whole, which exhibit more life, will have these fifteen properties to a strong degree. Conversely, the things in this world which are most lifeless will have these properties to the least degree.

1 Levels of Scale   2 Strong Centers   3 Boundaries   4 Alternating Repetition   5 Positive Space

6 Good Shape   7 Local Symmetries   8 Deep Interlock   9 Contrast   10 Graded Variation

11 Roughness   12 Echoes   13 The Void   14 Inner Calm   15 Not-Separateness

Global Mutation in Humanity

Russians seem to be ahead of the pack in their exploration of consciousness and the inter-relationships with scientific discoveries across various fields. We need not fear our fellow planetary citizens as they, too, are experiencing the inner revolution we are, too.

Recently my wife shared a video from a Russian Academician (Russia's Academy of Science highest title for scientists) named Valentina Mironova. She recited her dissertation on the Global Mutation in Humanity on a YouTube video. Fortunately it has English subtitles and, in spite of the usual boring monologue of reading dissertations live, it has some fascinating facts to correlate.

In essence, she shares that there is a cosmic configuration unfolding that affects humans at the DNA level. Apparently at least some of the 'junk DNA' is being triggered and a bio-spiritual evolutionary process is unfolding through our direct experience.

Conversations amongst scientists include revealing a fallacious narrative that has the effect of dividing our population to the point where those who are assigned to 'protect and serve' are displaying unconscionable behavior toward their fellow human beings. That alone ought to indicate a schism; cognitive dissonance to the extreme. The willingness to see that and begin to calm the storm with conscience and collaboration is the next task on a practical level, with the understanding of a deeper significance for all.

Perhaps by the time you read this we'll be progressing toward learning how to get along better and serve one another with the honor and respect we each deserve, in spite of the jabs and pokes. It is my hope that the material in this book has given you cause for a pause and reflection of the life you truly want to lead and how you can participate in the upwising of our planetary citizenry.

Excerpt from Mironova's Official Statement from 12.01.2020 *HUMAN.*

We all - together with the planet - make up a single whole. Of course, the event of cutting the Earth's polar ionosphere has already entered our flesh and blood. I would like to know how and where all this will lead.

If you think about it, then this whole picture turns out to be the most verified, calculated and duplicated that there is no doubt about the global scenario of the HIGHER MIND and His Highest goal.

The transitional phase is a comprehensive and global clean-up of the old energy plan. This conclusion suggests itself first. Because this PHENOMENON is immensely multifaceted and involves absolutely all levels of being. Some Birkeland currents are worth what - the very reconnection of the magnetic field lines, after which a new magnetic flow is established with all the new properties and field gradients. Including in the human body (brain).

MUTAGENIC EFFECT OF GAMMA RADIATION - or a new mechanism of DNA mutation

Under the influence of the flow of elementary particles coming from outer space, the molecular composition of DNA changes. The nucleotide forms a fast-living temporary compound that plays the role of a dynamic resector (dissector). Under the influence of an elementary particle, a "microexplosion" of the pentarhythmic ring occurs.

A fast-living resector compound is a nucleotide with the sixth member "H" (hydrogen) pulled into a ring break. By introducing new members, the DNA system loses its stability. But it gains mobility. This state lasts for an instantaneous fraction of a second (femto, atto and deeper). This is quite enough for memorizing the possibility of change and subsequent change.

The beginning of transmutation is always a loss of stability. The birth of 6, 7-component nucleotides marks the beginning of the process of mutations, where blood is a reflection of all mutations in the decompaction of the matter of the human body.

Actually, this structure is already gaining volatility and losing density. The principle of interaction of a chemical substance and a physical elementary particle provides unique opportunities for obtaining a new type of substance. And as a consequence - the birth of a new biochemistry (evolution and development) of a person to replace the past biochemistry (mandatory cellular stress).

In fact, this is already happening - the emergence of a "knot" type of DNA is an example of this; what was previously unknowingly called "junk" DNA.

And then one can only speculate.

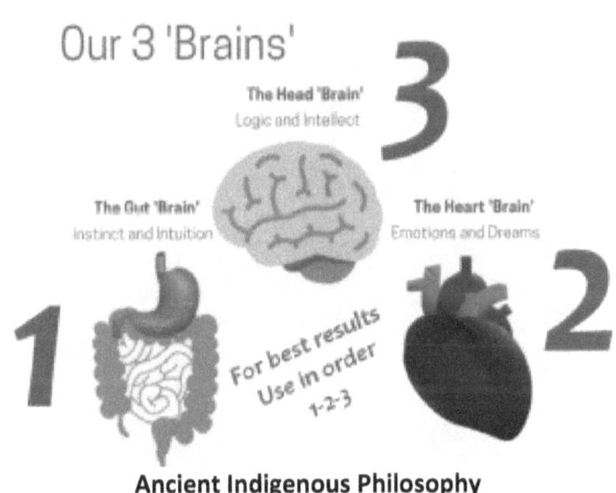

Ancient Indigenous Philosophy

*We cannot think
our way through
a system built on
vibration.
We have to FEEL
our way through it.*

Zen Benefiel

Conclusion of Sorts

We experience so many things inside, in multiple kinds of dimensions or worlds often foreign to us. We interpret them in so many ways, anecdotal to zeitgeist.

How do we respond and what are they really?

Those answers seem to be what we really seek, often unconsciously, and no doubt some with intention and volition. In the aspect of rising from a Type 0 to a Type I civilization with the expanded view, e.g. Michio Kaku and others have, on the Kardashev scale.

The movement toward a Type I civilization has many aspects that haven't been integrated, let alone identified, yet. The technological aspects of applied systems, human and material, toward effective planetary management are immense. That is the outer reality, what we might call self-actualized planetary administration. It is distinct and purposeful.

There is an inner perspective that, in all respects, emerges with insight, inspiration, investigation and personal identity as a conscious entity in a greater reality. This greater reality is a connected web of life, often driven by the awareness of abundant synchronicity, that is a result of the attention we put on intention and the actions we take in the spirit of 'acting as if.' This is where the inner and outer connect, quantum theory proven.

In theory, a natural order of mathematical precision could evolve into a living awareness of it. It is not a big leap in logic that a process would evolve to empower integration into the existing systems, like a holistic stewardship. That doesn't seem scary at all.

This kind of New World Order would seem to wear the title well, not appear as window dressing for a fear-based paradigm desperate to stay alive. As Buckminster Fuller said: "You never change things by fighting against the existing reality. To change something, build a new model that makes the old model obsolete."

With concerted effort, not bickering over resources, we could clean up the planet within a generation or less. If we include the Gaia Theory, which is congruent with the Quantum Theory as applied with attention, intention and action, the opportunity for a shift in weather patterns and possibly earthquakes to diminish might be present. Is it even possible for a planet and people to interact in such a conscious fashion? How would that need to happen? What would need to happen in order to observe that reality?

If you're one of those who like to go to the back of the book first, you are in luck. The theme of this book is 'harmony among people and planet' though it is presented with anecdotes, experiences, thoughtful contemplations and valuable definitions of contact modalities to cosmology to natural sensitivities often overlooked and certainly not understood. We headed toward the age of enlightenment just by the sheer amount of information we process.

Ufology has been trapped in old-world paradigms and we are in a new world order now, one that combines and synthesizes esoteric and scientific data that correlate and validate each other. There is a bridge that has been there for a long time. We're now just beginning to cross it, from fear to knowledge and understanding of the constructs of reality and our ability to perform in it.

I'll provide just a couple of references to the essence of the material, quotes from some thought-leaders still. Goethe's revealing, "Beauty is a manifestation of secret natural laws, which otherwise would have been hidden from us forever."

Montapert puts it very succinctly, "Every person has free choice. Free to obey or disobey the Natural Laws. Your choice determines the consequences. Nobody ever did, or ever will, escape the consequences of his choices."

Whether it be Cryptozoology, Psychology, Sociology or Ufology, we are evolving into a new living awareness of self and others. We've gone from tribal fiefdoms to a global village that, as of yet, has not acted like the holistic unit it is truly.

Emissaries from many worlds are attempting to get us to realize its truth. It is perhaps the only way we will unite and find ways to manage our human and planetary resources for any kind of sustainability on Earth.

How do we Be The Change? Simply, we become aware that attention, intention and action have a profound affect and effect on our environment and reality.

We make better choices when considering those in daily, even moment to moment, interactions with others and Nature. In effect, this experience actually proves the hypothesis of quantum physics – We create our reality.

Based on observation causing waves to become particles in the manifestation process, then it only stands to reason that we have a profound obligation. That obligation begins with gratitude for the opportunity of life. Live it fearlessly!

ZERO to ONE - Making Our Way Toward a Conscious Civilization is just an exploration of these considerations and how the indicators or signs might be present enough for us to be more aware and able to participate. According to ancient wisdom, the prudent path appears to those who are intent on finding it.

The truth is we are becoming aware of multiple components of a larger reality our belief systems and experience limited for millennia. The industrial age information age and increase in the availability of information catapulted web- connected humans into an accelerated learning curve with a parallel curve in the rise of awareness of inner and outer realities.

Perhaps now even the age of enlightenment is upon us and will bridge the gaps still present in various fields, from aerospace to zoology, and begin to form a holistic picture of the systems necessary to take us beyond the 21st Century.

Planetary sustainability requires us to learn how to work together cooperatively and collaboratively as a specie, as planetary citizenry and bio-social architects of a new living awareness. We advance from cultural and tribal constraints to a global society that recognizes diversity and appreciates uniqueness of individuals.

If/when we do not understand how to do so, we could indeed turn to civilizations that have done so and ask questions. We know they are here. We know they have been attempting to communicate in various ways in regard to this very issue – sustainability – and so far we've ignored them or we just are listening or looking in the right places to establish a line of communication.

Many hierarchical systems require a top-down communication and only the revelations from on high will suffice for distribution. This phenomenon is happening in reverse, apparently. People across various fields of study and personal experience are being approached and governments are not likely to listen to them until the leadership changes. Command and control is fading.

Collaboration and cooperation are on their way in, precipitating a new living awareness for moving forward. We are headed toward a civilization worthy of being part of a larger picture, or galactic perspective even though we are likely thousands of years from a Type III status.

The power of this change is truly within YOU and ME. Hopefully I've presented some indications that YES, there is a symmetry, a potential unity in the contact modalities and our normal waking consciousness. The two are inextricably connected.

Suggested Reading:

"The Field" by Lynne McTaggart - Explores the science of the interconnectedness of all things.

"Breaking the Habit of Being Yourself" by Dr. Joe Dispenza - Focuses on changing oneself through understanding the science of transformation.

"The Holographic Universe" by Michael Talbot - Discusses the theory that the universe is a hologram and its implications for understanding reality.

"Journeys Out of the Body" by Robert Monroe - Details personal experiences of out-of-body travel and its potential for personal growth.

"The Source Field Investigations" by David Wilcock - Examines a variety of topics related to science, consciousness, and ancient civilizations.

"The Cosmic Serpent" by Jeremy Narby - Explores the shamanic understanding of biology and its intersection with modern genetics.

"Fringe-ology" by Steve Volk - Investigates paranormal phenomena and their relation to human experience.

"The Conscious Universe" by Dean Radin - Focuses on the scientific evidence of psychic phenomena.

"Supernormal" by Dean Radin - Discusses the science of yoga and psychic abilities.

"DMT: The Spirit Molecule" by Dr. Rick Strassman - Explores the effects of DMT on consciousness and spiritual experiences.

Enjoy the journey, lead with vision and share great stories that inspire.

As a final gift, a sharing of conversations that matter and create a mindflow of similar direction and inclusive of the diverse experience –

https://youtube.com/@OneWorlinaNewWorld

Please join our subscriber list.

Let's learn how to work together:
PlanetaryCitizens.net